Medicinal plants in Bangladesh and its curative effects for disease

Mst Monira Khaton

Dr. Md Munan Shaik

Copyright © 2013 Mst Monira Khaton and Dr. Md Munan Shaik

All rights reserved.

ISBN: 149277880X
ISBN-13: 978-1492778806

DEDICATION

To Md Tahsin Shaik, our beloved son

Table of Contents

DEDICATION..iii

About the Book..ii

1. Chapter One: Introduction...1

2. Chapter Two: Important families..................................6

Fabaceae..6

Lamiaceae 10

Moraceae (Mulberry Family)...13

Solanaceae...15

Apocynaceae..18

Zingiberaceae 20

Euphorbiaceae...22

Rutaceae..25

Meliaceae...28

Combretaceae..29

Liliaceae 32

Umbelliferae...33

Araceae..34

Convolvulaceae..36

Asteraceae..38

Bromeliaceae...41

Others Families..42

3. Chapter Three: Medicinal Plants and Disease...........70

Cognitive enhancement..................................71

Rheumatoid arthritis.....................................72

Skin disease..73

Leprosy...74

Visceral leishmaniasis..................................75

Cytotoxic effect 75

Medicinal Plants and its Molecular effect.....................77

Antimirobial Activity......................................79

4 Chapter Four: In vitro Propagation and Conservation 84

Tissue culture...84

5. Chapter Five: Plants extracts and Metabolites.........94

6. Chapter six: Conservation...........................106

Biotechnology for the conservation of the Medicinal plants..108

7. Chapter seven: Registration & Property Rights......110

8. Chapter eight: RESEARCH AND DEVELOPMENT......112

9. Chapter nine: Conclusion...........................114

10. chapter ten: References..116

ABOUT THE AUTHOR...144

ABOUT THE BOOK

In traditional herbal medicine, numerous plants have been used to treat different diseases. In this review book we have compiled available published literatures and data about medicinal plants from different region of Bangladesh. In total of 231plants from around 50 different families are used for the treatment. The plants that are widely used mainly for folk medicine or aurdbadic treatment by triabal people, kavirajes and even by local peoples. The diseases that are mostly treated with the medicinal plants are very different: skin disease, common cold, fever, sexually transmitted disease, cancer, abdominal pain, rheumatism, gastric pain, women disease, and some of the different infectious disease. We also listed by the plants according to their families and the parts used for the treatment of diseases. We also summarized the diseases treated most with many differents plants. The metabolite extraction, in vitro propagation and conservation also discussed breifly. For some medicinal plants showed antimicrobial and cytotoxic activities are also presented.

1. CHAPTER ONE: INTRODUCTION

From the ancient times, plants have been used as medicines in everywhere of the world (Srivastava et.al., 1996). The exploitation of plants for therapeutic purposes increased at a surprising rate day by day (Walter, 2001; Rao and Arora, 2004). From 1970s, 21,000 medicinal plant species were used globally (WHO: Penso 1980; Anon. 2008a). In developing countries, 80% of people depend on medicinal plants to fulfill their primary health needs, occupying a key position on plant research and medicine (Ribeiro, 2010). Plant metabolites are an important source of medicines and contributing world health significantly (Lambert et. al., 1997, Hossain et al., 2003). Now a day, approximately 80% of traditional medicinal preparations involve the use of plants or plant extracts (Hopking et al., 2004). Bangladesh has a rich history of utilizing medicinal plants in the Ayurvedic, Unani, and the folk medicinal systems, which are practiced within the country (Rahmatullah et. al., 2010).

The use of plants as traditional medicine for the treatment of various diseases goes back a long time ago, around 3,000 BC (Sofowora, 1982). A number of contemporary pharmaceuticals have been derived from medicinal plants (Balick and Cox, 1996). By observing the curing methods of traditional medicine many important modern drugs have been derived includes, aspirin, atropine, ephedrine, digoxin, morphine, quinine, reserpine and tubocurarine (Gilani and Rahman, 2005). Bangladesh serves as a source of a wide variety of medicinal plants. Using traditional medicinal knowledge in drug discovery seems so promising that recently even large Pharmaceutical Companies have begin to show interest. Novartis, a Swiss drug giant, for example, has invested 100 million US dollars in a new research and development centre in Shanghai that will bring together modern biomedical research and millennia old medicinal concepts (Stone and Xin, 2006). It is also Novartis – in collabortaion with the world health organisation (WHO) – that produces and distributes Coartem® (Riamet®) a front line antimalarial drug. Coartem is derived from quinghausu (artemisinin) a sesquiterpene peroxide first isolated from the TCM herb *Artemisia annua* L. by Chinese scientists in 1971 (White, 2008).

About 2000 plants were recognized as medicinal plants in the 'materia medica' of traditional medicine in Southest Asian subcontinent and more than 500 of such medicinal plants have so far been enlisted as growing in Bangladesh (Ghani, 1998) in a total of 6,000 plant species occur in Bangladesh, of which about 300 are exotic and 8 are endemic. Of the total number of plant species 5000 are angiosperms, and 4 are gymnosperms. Ninety-five vascular plants have been rated as threatened, of which 92 are angiosperms, and 3 gymnosperms (IUCN, 2003; Banglapedia, 2006). Medicinal plants represent a rich source of antimicrobial agents. Plants are used medicinally in different countries and are a source of many potent and powerful drugs (Ghani, 1998).

A wide range of medicinal plant parts is used for extract as raw drugs and they possess varied medicinal properties. The different parts used include root, stem, flower, fruit, twigs exudate's and modified plant organs.

Over the past decade, there has been a dramatic increase in the demand for medicinal plants for use in Traditional Medicine (TM) and Contemporary and Alternative Medicine (CAM) in both developing and developed countries (Lee et al., 2008). In China, TM accounts for about 40% of all health care delivered (WHO 2002) and, in Amazonia, medicinal plants serve as the main form of health care for a majority of the populace (Shanley and Luz 2003).

Currently there is an urgent need to establish a checklist for Bangladeshi medicinal flora where all the families, species and infraspecific taxa should be arranged alphabetically. All the native species should be separated from the naturalized and adventitious plants. All the information regarding the medicinal plants published so far need to categorized based on the taxa, ethnomedical information/distribution, biological activities, and chemical data. In the Taxa, details such as Family, Latin name, vernacular name/s when known, synonyms, and an *Exsiccatum* needs to be specified. For chemical data and biological activity tested *in vitro* and *in vivo*, all the literature published so far in journal need to collect and summarized based on the different keywords for proper further used of this valuable studies. But practically no effort was found for Bangladeshi medicinal flora, except "The Ethnobotanical Database of Bangladesh (EDB)" has compiled, which contains on line plant database in Bangladesh. In the current study all the medicinal plant reported to be found in Bangladesh are categorized based on the family with the description of their medicinal value for the selected disease and the plant parts used. *In vitro* propagation, selective metabolites extraction and their properties, intellectual property right along with the future prospects also described.

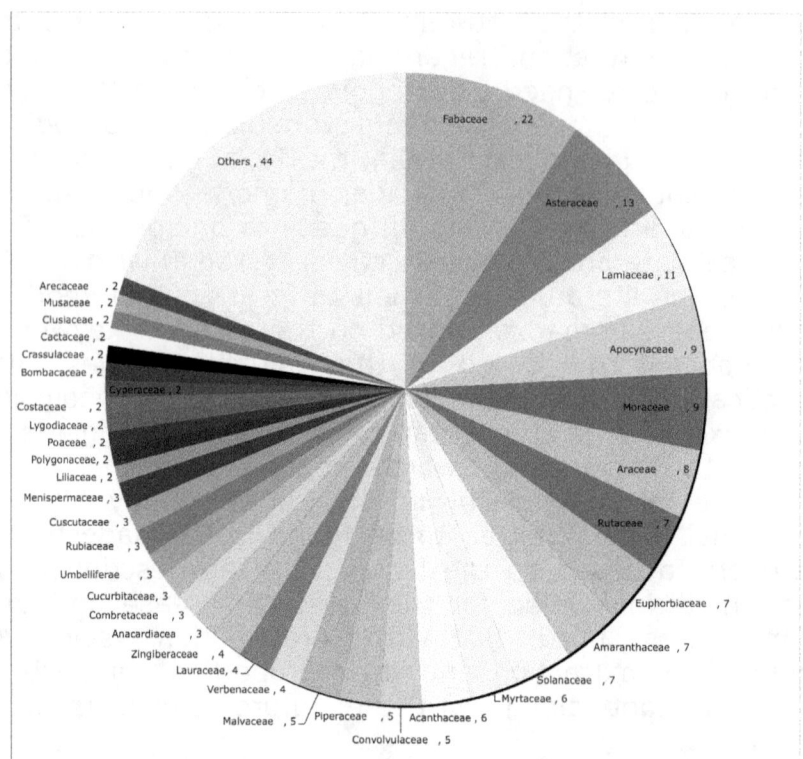

Figure 1. 37 families are reported to be presents the major portion of the medicinal plant in Bangladesh. The number of species in each families are shown.

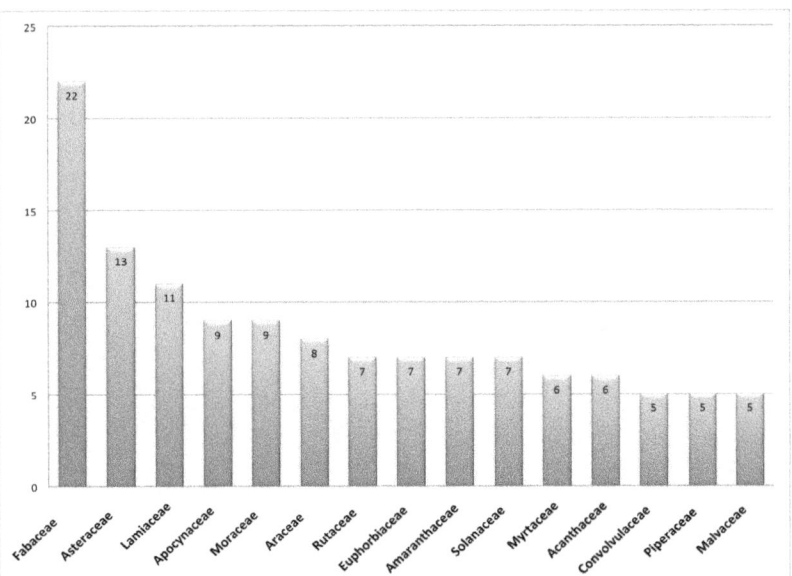

Figure 2: The most reported medicinal plants families in Bangladesh and the number of species in each families reported in literature.

Khaton and Shaik

2. CHAPTER TWO: IMPORTANT FAMILIES

Fabaceae

Over 490 medicinal plant species makes the Fabaceae as the second largest family in medicinal plants, most of which have been used as traditional medicines. There are 31 species of medicinal plants belonging to 20 genera in the family Fabaceae identified in the Chinese Pharmacopoeia (Chinese Pharmacopoeia Commission, 2005), as well as numerous species that are included in the Japanese Pharmacopoeia. Additionally, plant materials from nearly 290 species belonging to 100 genera of Fabaceae have been reported to be toxic (Gao *et al.*, 2010). In Bangladesh we found twenty species used as traditional medicine for the treatment of different diseases (Table 1). The plant leaves, bark and seed of Sishu (*Dalbergia sissoo Roxb. Ex*) is used to treated skin disease such as Eczema and sexual disease (Rahman *et. al.*, 2008); plant leaves, root and flower of Lajjaboti (*Mimosa pudica L.*) is used for the treatment of Diarrhea, hypertension, antidote to Poison (Rahman *et. al.*, 2008; Uddin *et. al.*, 2006); plant root of Neel aparajita*(Clitoria ternatea* L.) has been used for the treatment Stomachache problem (Mukherjee, 2008; Melanie-Jayne et. al., 2003; Rahmatullah et. al., 2010b) as a memory enhancer, nootropic, antistress, anxiolytic, antidepressant, anticonvulsant, tranquilizing and sedative agent (Misra, 1998; Warrier et al., 1995).

Table 1. Plants identified from the family *Fabaceae* with their uses and local name.

Species	Local name	Parts used	Aliments	References
Dalbergia sissoo Roxb. ex DC.	Shishu	Leaves, bark, seed	Eczema, sexual diseases,	Rahman et. al., 2008; Rahmatullah et. al., 2010c; Rahmatullah et. al., 2010e
Mimosa pudica L.	Lajjaboti	Leaves, root, flower	Diarrhea, hypertension, antidote to poison.	Rana et. al., 2010; Rahman et. al., 2008; Uddin et. al., 2006; Rahmatullah et. al., 2009; Ling et. al., 2009; Rahmatullah et. al., 2010c; Hossan et. al., 2010; Rahmatullah et. al., 2010e
Saraca indica L.	Ashok	Leaves, bark	Sexual diseases, analgesic, appetizer.	Rahmatullah et. al., 2009; Tewari et. al., 2000; Rahmatullah et. al., 2010e; Khan & Rashid, 2006
Cassia fistula L.	Bandor lathi	Root	Anal disorders (prolapse). Roots of theplant are mixed with roots of Mesua ferrea, crushed	Rana et. al., 2010; Ling et. al., 2009; Rahmatullah et. al., 2010b; Chowdhury et. al., 2010; Mukul et. al., 2007

Medicinal plants in Bangladesh and its curative effects for disease

Species	Local name	Parts used	Aliments	References
			and taken.	
Clitoria ternatea L.	Neel aparajita	Root	Stomachache	Mukherjee PK, 2008; Melanie-Jayne et. al., 2003; Rahmatullah et. al., 2010b; Rahmatullah et. al., 2010c; Rahmatullah et. al., 2010d; Hossan et. al., 2010;
Mucuna pruriens (L.) DC.	Alkushie	Seed	Increase sperm count, stimulate energy, wounds to vaginal area in women	Rahmatullah et. al., 2010b
Cajanus cajan	Arhar	Root, Leaves	Juice made from young leaves are used in jaundice and juice made from roots are used in diabetes	Rahman et. al., 2008; Rahmatullah et. al., 2009; Rahmatullah et. al., 2010c
Cassia fistula L.	Shonail, Bandor lathi	Fruit	Head infections. Fruit juice is applied to infected areas	Rahman a et. al., 2010; Chowdhury et. al., 2010
Cassia tora L	Henchai	Seed	Sex	Rahman a et. al.,

Species	Local name	Parts used	Aliments	References
			stimulant	2010
Mimosa diplotricha C. Wright	Shada lojjaboti	Root	Headache, pain in the forehead	Rana *et. al.*, 2010; Rahman *et. al.*, 2008; Uddin *et. al.*, 2006; Rahman a *et. al.*, 2010
Cassia fistula L	Shonail, Bandor lathi	Top portion of young stems	Purgative	Rahman a *et. al.*, 2010; Rahmatullah *et. al.*, 2010c
Erythrina variegata L.	Madar gach	Leaves, bark	Lower abdominal pain, diarrhea, Blood clotting	Rahman *et. al.*, 2008; Rahmatullah *et. al.*, 2010c; Chowdhury *et. al.*, 2010; Rahmatullah *et. al.*, 2010d
Lathyrus sativus L.	Khesari dal	Seed	Scabies, eczema, allergy	Chowdhury *et. al.*, 2010
Abrus precatorius L.	Shet kunch	Leaves, root	Aphrodisiac, arthritis, rheumatism	Rahmatullah *et. al.*, 2010d
Cassia alata L.	Daoder gach	Leaves	Ringworm	Rahmatullah *et. al.*, 2010d
Desmodium gangeticum (L.) DC.	Shalpan	Leaves, root	Chest pain, sexual problems	Rahmatullah *et. al.*, 2010d
Tamarindus indica L.	Tetul gach	Leaves, flower	Dysentery, Eye disease	Rahman *et. al.*, 2008;

Medicinal plants in Bangladesh and its curative effects for disease

Species	Local name	Parts used	Aliments	References
			, cataract, Rheumatism	Rahmatullah et. al., 2010c, ; Rahmatullah et.al, 2010d; Rahmatullah et. al., 2010e
Albizia lebbeck (L.) Benth.	Shirish	Leaves, root, bark, fruit	Asthma, coughs, thyroiditis, night blindness, diabetes, toothache, insect and animal bite	Rahmatullah et. al., 2010c
Erythrina variegata L.	Paldi madar	Leaves juice	Head lice infestation, lice infestation in animals	Rahmatullah et. al., 2010c
Albizia procera Benth	Koroi	Leaves	Allergy	Uddin et. al., 2006

Lamiaceae

The largest number of aromatic plants consist the Lamiaceae family, mainly used in aromatherapy. The essential oils extracted from the plants used to treat different disease. There are eleven species idenfied in Bangladesh used as medicinal plant. Among then leaves of tulshi (*Ocimum gratissimum* L.) used for the treatment of heart diseases, eye diseases, coold diseases, bronchitis, liver diseases, cancer (Rana *et. al.*, 2010; Uddin *et. al.*, 2006; Mukul *et. al.*, 2007; Rahmatullah *et. al.*, 2010e); root and bark of Bamon hati (*Clerodendrum indicum* L.) used for the treatment of asthma, chicken pox in children (Rahmatullah *et. al.*, 2009; Rahmatullah *et. al.*, 2010c).

Table 2. Plants identified from the family *Lamiaceae* with their uses and local name.

Species	Local name	Parts used	Aliments	References
Ocimum gratissimum L.	Seth-tulshi	Whole plant	Heart diseases, eye diseases, cooling	Rana *et. al.*, 2010; Rahmatullah *et. al.*, 2010e
Ocimum tenuiflorum L.	Krishna-tulshi	Whole plant	Bronchitis, liver diseases, cancer	Rana *et. al.*, 2010; Uddin *et. al.*, 2006; Mukul *et. al.*, 2007; Rahmatullah *et. al.*, 2010e
Leonurus sibiricus L.	Rokto-dron	Whole plant	Blood clotting	Rahman a *et. al.*, 2010
Leucas aspera (Willd.) Link	Kaun shim	Leaves	Sudden feeling of warmth head, headache	Rahman a *et. al.*, 2010

Medicinal plants in Bangladesh and its curative effects for disease

Species	Local name	Parts used	Aliments	References
Ocimum gratissimum L.	Tulshi	Leaves	Coughs, mucus	Khan *et. al.*, 2006; Rana *et. al.*, 2010; Rahman *et. al.*, 2008; Uddin *et. al.*, 2006; Rahman a *et. al.*, 2010; Rahmatullah *et. al.*, 2010b; Rahmatullah *et. al.*, 2010c; Rahmatullah *et. al.*, 2010d; Chowdhury *et. al.*, 2009; Khan & Rashid, 2006
Clerodendrum Indicum (L.) Kuntze	Bain josthi	Top of stem Without leaves	Eczema, itches	Chowdhury *et. al.*, 2010
Leonurussibiricus L.	Jongli bhang	Whole plant	Good sleep	Chowdhury *et. al.*, 2010; Rahmatullah *et. al.*, 2010e
Leonurus sibiricus L	Seth-dron	Leaves, flower	Mucus, helminthiasis	Rahmatullah *et. al.*, 2009
Leucas aspera (Willd.)	Dondo-kolosh	Leaves, flower	antipyretic, antirheumatic, anti-inflammatory, and antibacterial treatment,	Rahmatullah *et. al.*, 2009; Ahmed et.al.,2010
Clerodendru	Bamon	Root,	Asthma,	Rahmatullah *et.*

Species	Local name	Parts used	Aliments	References
m indicum (L.) Kuntze	hati	bark	chicken pox in children	*al.*, 2009; Rahmatullah *et. al.*, 2010c
Menthe arvensis	Pudina	whole plant.	Dysentery, indigestion, stomach pain.	Rahmatullah *et. al.*, 2009

Moraceae (Mulberry Family)

The mulberry family occurs primarily in tropical and semi-tropical regions, and includes a wide variety of herbs, shrubs, and trees, characterized by a milky sap and reduced, unisexual flowers. This family includes 40 genera and 1,000 species worldwide, but in Bangladesh only seven species recognized as used in traditional medicine. Mainly dumur (*Ficus hispida* L.f.) is used for the treatment of different disease such as for the internal bleeding stomach ache (Uddin et. al., 2006; Mollik et. al., 2010;); leaves, root, gum of Shaowra (*Streblus asper* L) for constipation, skin disorder, indigestion (Mollik et. al., 2010; Rahmatullah et. al., 2009; Chowdhury et. al., 2010; Rahman et. al., 2010); areal root of bot (*Ficus benghalensis* L.) for low semen density, diabetes (Rahman et. al., 2010; Rahman et. al., 2008; Mollik et. al., 2010).

Table 3. Plants identified from the family Moraceae with their uses and local name.

Species	Local name	Parts used	Aliments	References
Ficus hispida L.	Kack dumur	Whole plant	Internal bleeding	Uddin et. al., 2006; Mollik et. al., 2010; Rahmatullah et. al., 2010b; Chowdhury et. al., 2009
Artocarpus heterophyllus Lam.	Kathal	Fruit, sap	Vitamin supplementation	Rahman et. al., 2008; Rahmatullah et. al., 2010d ; Chowdhury et. al., 2009
Artocarpus lacucha Buch.-Ham.		Leaves, Fruit pericarp		Hasan et. al., 2009a
Ficus racemosa L.	Jog dumur	Fruit	Stomach ache	Uddin et. al., 2006; Mollik et. al., 2010; Chowdhury et. al., 2010; Rahmatullah et. al., 2010d
Ficus hispida L.	Khoskhose goj bukhil	Root	Dysentery, Diabetes	Mollik et. al., 2010; Chowdhury et. al., 2010
Streblus asper Lour.	Shaowra	Leaves, root, gum	Constipation, skin disorder. Indigestion	Mollik et. al., 2010; Rahmatullah et. al., 2009; Chowdhury et. al., 2010; Rahman et. al.,

Species	Local name	Parts used	Aliments	References
				2010
Ficus benghalensis L.	Bot	Root (aerial root)	Low semen density, diabetes	Rahman et. al.,2010; Rahman et. al., 2008; Mollik et. al., 2010; Rahmatullah et. al., 2009; Chowdhury et. al., 2010; Rahmatullah et. al., 2010d

Solanaceae

Solanaceae includes over 3,000 species of plants is extensively utilized by humans as source for food, spice, and medicine, and ornamentals. Members of Solanaceae provide stimulants, poisons, narcotics, pain relievers, and so forth and often rich in alkaloids. From Solanaceae family we found seven species in Bangladesh reported as medicinal plant. Leaves, root, seed of Dhutra (*Datura metel* L.) is used for the treatment of swelling and Pain, breast pain, anesthesia, asthma, epilepsy, rheumatic fever, hypertension (Rahman et. al., 2010; Khan et. al., 2006; Rahmatullah et. al., 2010b); seeds of tit baegun (*Solanum Capsicoides* Allioni) for Snake bite (Ling et. al., 2009; Rahmatullah et. al., 2010c; Rahmatullah et. al., 2009; Chowdhury et. al., 2010); root of ashwogondha

(*Withania somnifera* L.) for piles, desbility and cancer (Tewari et. al., 2000; Rahmatullah et. al., 2010b; Rahmatullah et. al., 2010d), age associated decline in cognitive function (Parrotta, 2001).

Table 4. Plants identified from the family Solanaceae with their uses and local name.

Species	Local name	Parts used	Aliments	References
Datura metel L	Dhutura	Leaves, root, seed	Swelling and Pain, excessive breathing, enlarge pupil in eye, swelling of gums and base of ears, breast pain. anesthesia, pain, asthma, epilepsy, rheumatic fever, hypertension	Rahman et. al., 2010; Khan et. al., 2006; Rahmatullah et. al., 2010b; Rahmatullah et. al., 2010c; Rahman et. al., 2008; Rahmatullah et. al., 2009; Rahmatullah et. al., 2010d
Withania somnifera (L.) Dunal	Ashwogondha	Root	Piles, debility. Cancer	Sharmen et. al., 2005; Tewari et. al., 2000; Ven Murthy MR, 2010 SK Kulkarni - 2008; Rahmatullah et. al., 2010b; Rahmatullah et.

Species	Local name	Parts used	Aliments	References
				al., 2010d
Solanum capsicoides Allioni	Kontikari, Sial kata	Whole plant	Low semen count, itches. endocrinological, disorders, diabetes	Rahman et. al., 2010; Rahmatullah et. al., 2010c; Chowdhury et. al., 2010; Rahmatullah et. al., 2010d
Datura metel L.	Shada dhutura	Leaves	Scabies, eczema, allergy	Rahman et. al., 2008; Chowdhury et. al., 2010
Solanum capsicoides Allioni	Tit baegun	Seed	Snake bite	Ling et. al., 2009; Rahmatullah et. al., 2010c; Rahmatullah et. al., 2009; Chowdhury et. al., 2010
Solanum indicum L.	Khelna	Leaves	Hypertension, poisonous insect bites	Rahmatullah et. al., 2009
Physalis micrantha Link	Tulo tepa	Leaves	Tonsillitis	Rahmatullah et. al., 2010d

Apocynaceae

This family consists of several important medicinal plants with wide range of pharmacological, biological acitivies. Among the ten identified medicinal species in Bnagladesh root of shorpogondha (*Rauwolfia serpentine* L.) used for the treatment of stomach pain (Khan *et. al.*, 2006; Rahman *et. al.*, 2008; Rahmatullah *et. al.*, 2009;); bark of chatim (*Alstonia scholaris* L.) used for the treatment of continuous fever, malaria (Rana *et. al.*, 2010; Rahman *et. al.*, 2008; Rahmatullah *et. al.*, 2009; Chowdhury *et. al.*, 2010; Mukul *et. al.*, 2007; Chowdhury *et. al.*, 2009).

Table 5. Plants identified from the family Apocynaceae with their uses and local name.

Species	Local name	Parts used	Aliments	References
Hemidesmus indicus (L.) R.Br. ex Schult.	Anantamul	Leaves	Urinary tract infections	Hossan *et. al.*, 2010; Rahmatullah *et. al.*, 2010d
Holarrhena pubescens Buch.-Ham. Wall. ex G. Don	Koruj gach	Bark	Leucorrhoea	Hossan *et. al.*, 2010; Rahmatullah *et. al.*, 2010b,
Rauwolfia serpentina (L.) Benth.ex Kurz	Shorpogondha	Root	Stomach pain.	Khan *et. al.*, 2006; Rahman *et. al.*, 2008; Rahmatullah *et. al.*, 2009; Tewari *et. al.*, 2000; Rahman a *et. al.*, 2010; Khan & Rashid, 2006
Tabernaemo	Kori	Flower	Conjunctivit	Rahman a *et.*

Species	Local name	Parts used	Aliments	References
ntana (L.) R. Br. Ex Roemer & J.A. Schultes			is. Eye disease	al., 2010
Alstonia scholaris (L.) R.Br.	Chatim	Bark	Continuous fever, malaria.	Rana et. al., 2010; Rahman et. al., 2008; Rahmatullah et. al., 2009; Chowdhury et. al., 2010; Mukul et. al., 2007; Chowdhury et. al., 2009
Thevetia peruviana (Pers.) K. Schum.	Mon shila, Yellow flowere Kolki phool	Fruit	Rheumatism	Chowdhury et. al., 2010
Allamanda cathartica L.	Koilki phool gach	Root	To provide cooling effect in body.	Rahmatullah et. al., 2009; Rahmatullah et. al., 2010d
Tabernaemontana divaricata (L.)	Kath mollika	Leaves	Sedative	Rahmatullah et. al., 2009; Rahmatullah et. al., 2010d

Medicinal plants in Bangladesh and its curative effects for disease

Species	Local name	Parts used	Aliments	References
Catharanthu sroseus (L.) G.Don	Noyon-tara	Leaves, flower	Diabetes, helminthiasis	Rahmutullah et. al., 2009; Ling et. al., 2009; Rahmatullah et. al., 2010c
Vallaris solanaceae Kuntze (syn.: V. heynei)	Agarmoni	Flowers	ringworms and skin infections.	Ahmed et.al.,2010

Zingiberaceae

The Zingiberaceous plants are characterized by their tuberous or non-tuberous rhizomes, which have strong aromatic and medicinal properties and exists in about 1,300 species worldwide, distributed mainly in South and Southeast Asia (Wu and Larsen, 2000). In Bangladesh we found only five species are used for medicinal purposes. Six species of the Zingiberaceae family (*Zingiber cassumunar, Alpinia galanga, Alpinia allughas, Hydychium corancium, Hydychium coccinum* and *Kaempferia galaugalim*) shown to potential antioxidant activity (Vankar, 2006)

Turmeric (*Curcuma longa* L.) is used in ancient Indian medicine and in cosmetics, to help clean and heal wounds, to treat asthma, dysmenorrhoea, psoriasis, eczema, arthritis, hepatic and digestive disorders and to prevent cardiovascular diseases, to reduce cancer risk (Sakarkar et al., 2006), to counteract ageing processes

(Yu et. al., 2002). Ginger (*Zingiber officinale*) has antibacterial, antifungal, antiparasitic, antiviral, anti-diabetic, anti-inflammatory, antioxidant and anti-hypercholesterolaemic properties (Saeed and Tariq, 2006). Ginger to be especially effective in curbing motion sickness, morning sickness, and postoperative and chemotherapy-induced nausea. Ginger is an excellent digestive, aiding in the absorbtion of food and elimination of gas and bloating. Ginger stimulates circulation so it is good for cold hands and feet.

Table 6. Plants identified from the family Zingiberaceae with their uses and local name.

Species	Local name	Parts used	Aliments	References
Zingiber officinale Rosc.	Ada	Rhyzome	Cold, Cough	Sharmen et. al., 2005; Bhowmic et. al., 2008; Uddin et. al., 2006; Ling et. al., 2009; Melanie-Jayne et. al., 2003; Mollik et. al., 2010; Rahmatullah et. al., 2010d; Patil and Bhaskar, 2006 ; Mukul et. al., 2007; Chowdhury et. al., 2009; Khan & Rashid, 2006
Curcuma	Kalo	Stem	Pain,	Khan et. al.,

Medicinal plants in Bangladesh and its curative effects for disease

Species	Local name	Parts used	Aliments	References
caesia Roxb.	holud		rheumatism, insanity, uterus infections, diarrheoa	2006; Rahman a et. al., 2010; Rahmatullah et. al., 2010b; Rahmatullah et. al., 2010d; Khan & Rashid, 2006
Curcuma zedoaria (Christm.) Roscoe	Shoti gach	Leaves	Diabetes, constipation, indigestion	Rahman a et. al., 2010; Rahmatullah et. al., 2010c; Rahmatullah et. al., 2010d
Curcuma alismatifolia Gangnep.		Leaves		Hasan et. al., 2009°; Akter et. al., 2008
Curcuma longa Linn.	Holud	Rhizome	Skin diseases	Mukul et. al., 2007; Chowdhury et. al., 2009; Rahmatullah et. al., 2010d; Khan & Rashid, 2006

Euphorbiaceae

Euphorbiaceae is a large family of flowering plants around 7,500 species (Gill, 1988). The family provides food (Pandey, 2006; Etukudo, 2003) and varied medicinal properties used in ethnobotany (Gill, 1988; Vasishta, 1974; Agbovie et al., 2002; Betti, 2004; Kubmarawa, 2007) and used in the treatment of ailments such as respiratory infections, venereal diseases, toothache, rheumatism, cough, ulcer and wounds (Oliver, 1960). Seven species found in Bangladesh used for medicinal purpose. Amla (*Emblica officinalis*) have anticholinesterase activity and it is also used to lower the blood cholesterol level (Vasudevan and Parle, 2007); plants leaves, fruit of amloki (*Phyllanthus emblica* L.) is used for the treatment of the appetizer, gonorrhea, toothache, itch, leucorrhoea (Bhowmik et. al., 2008; Tewari et. al., 2000; Khan et. al., 2006; Rahman et. al.,2008; Uddin et. al., 2006; Rahmatullah et. al., 2009; Rahman a et. al., 2010), Leaves and root of Muktarjhuri (*Acalypha indica* L.) for aphrodisiac, to ease delivery, passing of semen with urine (Rahmatullah et. al., 2010c; M Rahmatullah et. al., 2010d; Rahmatullah et. al., 2010e), Laal kocha (*Jatropha gossypifolia* L.) reported to have anticancer, hepatoprotective and pesticidal activity (Shahwar, 2010), sap is used for the treatment Abscess in hip (Ling et. al., 2009; Rahmatullah et. al., 2010d), leaf is used for bathing wounds (Sosa et al., 2002; Kayaalp, 1998; Hartwell, 1969; Chatterjee et al., 1980; Panda et al., 2009).Veron (*Ricinus communis*) known as castor oil tree, is widely used as a human laxative-cathartic agent, (Fingl, 1980), whole and seeds used for constipation, arthritis, rheumatism, atomachache, bloating, decreased eyesight conjunctiviti (Chowdhury et. al., 2010; Ling et. al., 2009;

Rahmatullah *et. al.*, 2009; Rahmatullah *et. al.*, 2010d). *Emblica officinalis* is a plant with rasayana properties and used for the preparation of tonic as a main ingredient for a disease-free life with long lasting youth, great vigour and no dementia (Bala and Manyam, 1999).

Table 7. Plants identified from the family Euphorbiaceae with their uses and local name.

Species	Local name	Parts used	Aliments	References
Emblica officinalis Gaertn.	Amla	Leaves, fruit	Leucorrhoea	Bhowmik *et. al.*, 2008; Tewari *et. al.*, 2000; Hossan *et. al.*, 2010;
Phyllanthus emblica L.	Amloki	Leaves, fruit	Appetizer, gonorrhea, toothache, itch.	Khan *et. al.*, 2006; Rahman *et. al.*,2008; Uddin *et. al.*, 2006; Rahmatullah *et. al.*, 2009; Rahman a *et. al.*, 2010; Rahmatullah *et. al.*, 2010c; Chowdhury *et. al.*, 2010; Mukul *et. al.*, 2007; Chowdhury *et. al.*, 2009; Rahmatullah *et. al.*, 2010e
Ricinus communis L.	Veron	Oil from whole plant,	Constipation. Arthritis, rheumatism	Chowdhury *et. al.*, 2010; Ling *et. al.*, 2009; Rahmatullah *et.*

Species	Local name	Parts used	Aliments	References
		Leaves, seed	Stomachache, bloating, decreased eyesight conjunctiviti	al., 2009; Rahmatullah et. al., 2010d
Acalypha indica L.	Muktarjhuri	Leaves, root	Aphrodisiac, to ease delivery, passing of semen with urine.	Rahmatullah et. al., 2010c; M Rahmatullah et. al., 2010d; Rahmatullah et. al., 2010e
Jatropha gossypifolia L.	Laal kocha	Sap	Abscess in hip	Ling et. al., 2009; Rahmatullah et. al., 2010d
Euphorbia milii 'Lutea' Hort	Dodhi-kata	Whole plant	Eczema, sexual diseases, diarrhea.	Rahmatullah et. al., 2010e

Rutaceae

The Rutaceae family is a small family that consists of about 1300-1600 species throughout the world contain cultivated fruit trees and medicinal herbs. In Bangladesh eight species reported to be used as medicinal plant. Bel fruit (*Aegle marmelos*(L.) is used for the treatment of heat stroke, ulcer, Indigestion (Rahman et. al., 2008; Tewari et. al., 2000; Melanie-Jayne et. al., 2003; Rahmatullah et. al., 2010b). Roots of Jambura gach (*Citrus grandis* L.) is used for the treatment of aphrodisiac (Melanie-Jayne et. al., 2003; Rahmatullah et. al., 2010d).

Table 8. Plants identified from the family Rutaceae with their uses and local name.

Species	Local name	Parts used	Aliments	References
Aegle marmelos (L.) Correa.	Bel	Fruit	Heat stroke, Ulcer, Indigestion,	Sharmen et. al., 2005; Rahman et. al., 2008; Tewari et. al., 2000; Melanie-Jayne et. al., 2003; Rahmatullah et. al., 2010b; Rahmatullah et. al., 2010c; Rahmatullah et. al., 2010d; Patil and Bhaskar, 2006; Mukul et. al., 2007; Chowdhury et. al., 2009
Citrus aurantium L.	Komla	Fruit skin	Stomach pain	Melanie-Jayne et. al., 2003; Chowdhury et. al., 2009
Glycosmis pentaphylla (Retz.) Corr.	Dontomul, Atkhira, Atali gach	Leaves, root	Frequent urination, low sperm count, Enlarged	Melanie-Jayne et. al., 2003; Rahman a et. al., 2010; Chowdhury et.

Species	Local name	Parts used	Aliments	References
			liver	al., 2010
Murraya paniculata (L.) Jack	Kamini	Leaves	Dysentery	Rahmatullah et. al., 2009; Melanie-Jayne et. al., 2003; Rahmatullah et. al., 2010b; Chowdhury et. al., 2010
Citrus aurantifolia (Chris. & Pan.) Sw.	Lebu gach	Fruit	Sexual stimulant	Ling et. al., 2009; Rahmatullah et. al., 2010b; Rahmatullah et. al., 2010d ; Chowdhury et. al., 2009
Citrus grandis (L.) Osbeck	Jambura gach	Root	Aphrodisiac	Melanie-Jayne et. al., 2003; Rahmatullah et. al., 2010d
Glycosmis pentaphylla (Retz.) Corr	Gondho-pata Mootkila	Leaves	Diabetes, to prevent termite damage	Rahmatullah et. al., 2009
Feronia limolia Linn.		Leaves		Hasan et. al., 2009a

Meliaceae

Four species from the family Meliaceae was reported to be used for traditional treatment in Bangladesh. Neem (*Azadirachta indica*) is a useful traditional medicinal plant and most widely used and also every parts is commercially exploitable for the development of medicines and industrial by-products. Leaves, bark, seed is used for the treatment of different disease such as, cancer, skin diseases, helminthiasis, wounds, diabetes, rheumatoid arthritis, scabies, eczema, itches (Khan *et. al.*, 2006; Rana *et. al.*, 2010; Rahman *et. al.*, 2008; Bhowmik *et. al.*, 2008; Uddin *et. al.*, 2006). Neem tree reported to work against bacteria, viruses, fungi, lice, and even cancer. It may also prevent pregnancy. Leaf, bark, seed of Mahogany (*Swietenia mahagoni* L.) is used for the treatment of impotency, malaria, appetizer (Rahmatullah *et. al.*, 2010e), anti platelet and anti microbial activity (Ekimoto et. al., 1991; Rahman et. al., 2009).

Table 9. Plants identified from the family Meliaceae with their uses and local name.

Species	Local name	Parts used	Aliments	References
Azadirachta indica A. Juss.	Neem	Leaves, bark	Cancer, skin diseases, helminthiasis, wounds, diabetes, rheumatoid arthritis, Scabies, eczema, itches.	Khan *et. al.*, 2006; Rana *et. al.*, 2010; Rahman *et. al.*, 2008; Bhowmik *et. al.*, 2008; Uddin *et. al.*, 2006; Ling *et. al.*, 2009; Mollik *et. al.*, 2010; Rahman a *et. al.*, 2010;

Species	Local name	Parts used	Aliments	References
				Rahmatullah et. al., 2010b; Chowdhury et. al., 2010; Rahmatullah et. al., 2010d ; Mukul et. al., 2007; Chowdhury et. al., 2009; Rahmatullah et. al., 2010e; Khan & Rashid, 2006
Aphanamixis polystachya (Wall.) R. Parker	Pitraj	Leaf, bark, seed	Rheumatoid arthritis, analgesic, itch, antidote to poison.	Rahmatullah et. al., 2010e
Swietenia mahagoni (L.) Jacq.	Mahogany	Leaf, bark, seed	Impotency, malaria, appetizer.	Rahmatullah et. al., 2010e
Cereus grandiflorus (L.) P.Mill.	Kuth-raaz	Whole plant	Tonic, nervous disorders, heart diseases.	Rahmatullah et. al., 2010e

Combretaceae

Combretaceae family includes about 600 species of which three species are reported in Bangladesh used in Traditional medicine. Arjun (*Terminalia arjuna*) is used in the treatment of glositis, chelitis and gingivitis (Ghani, 2000). The fruits of haritaki (*Terminalia chebula*) have

Medicinal plants in Bangladesh and its curative effects for disease

different medicinal effectsincluding, treatment of indigestion, constipation, dysentery, jaundice, piles, dysmenorrhea. lacrimation, chronic ulcers and wounds (Chopra et al., 1956). The ripe fruit of haritaki is regarded as a promoter of intellect and memory, and is believed to retard the ageing process and to improve cognitive processes (Manyam, 1999; Misra, 1998; Vohra and Gupta, 2005). Bohera (*Terminalia belerica*) fruits are commonly and largely used in indigenous medical purposes. The pulp of the fruit is used in treating diarrhoea and leprosy. Half-ripped fruits are used as purgative. The oil extracted from the seeds is useful as a hair tonic.

Table 10. Plants identified from the family Combretaceae with their uses and local name.

Species	Local name	Parts used	Aliments	References
Terminalia arjuna (Roxb.) Wight & Arn.	Arjun	Leaves, bark, fruit	Hypertension, Anemia, Leprosy, Low sperm count, dysentery. Heart disorders, indigestion	Khan *et. al.*, 2006; Rana *et. al.*, 2010; Rahmatullah *et. al.*, 2010b; Rahman *et. al.*, 2008; Bhowmik *et. al.*, 2008; Uddin *et. al.*, 2006; Rahman a *et. al.*, 2010; Rahmatullah *et. al.*, 2010c; Chowdhury *et. al.*, 2010;

Species	Local name	Parts used	Aliments	References
				Rahmatullah et. al., 2010d; Mukul et. al., 2007; Chowdhury et. al., 2009 ; Rahmatullah et. al., 2010e; Khan & Rashid, 2006
Terminalia belerica (Roxb.)	Bohera	Leaves, bark,	sexual diseases	Rana et. al., 2010; Uddin et. al., 2006; Rahman a et. al., 2010; Mukul et. al., 2007; Chowdhury et. al., 2009; Rahmatullah et. al., 2010e; Khan & Rashid, 2006
Terminalia chebula Retz.	Horitoki	Fruit	Chronic mucus, Flu, skin allergy	Rana et. al., 2010; Rahman et. al., 2008; Bhowmik et. al., 2008; Rahmatullah et. al., 2010b; Uddin et. al., 2006; Melanie-Jayne et. al., 2003; Rahmatullah et. al., 2010c; Chowdhury et. al., 2010; Mukul et. al., 2007; Chowdhury et. al., 2009; Rahmatullah et. al., 2010e; Khan

Medicinal plants in Bangladesh and its curative effects for disease

Species	Local name	Parts used	Aliments	References
				& Rashid, 2006

Liliaceae

The Liliaceae or lily family is composed of large number of plant with medicinal virtues. Among the two species of plant reported from this family used in Bangladesh for medicinal purposes. Leaves and root of shotomul (*Asparagus racemosus* Willd.) used for the treatment for diabetes, heart diseases (Rahman *et. al.*, 2008; Rahmatullah *et. al.*, 2009; Tewari *et. al.*, 2000), leaves go-roshun (*Crinum asiaticum* L.) used for treated dysentery in cattle (Rahmatullah *et. al.*, 2009; Ling *et. al.*, 2009), roots of *Asparagus racemosus* are used against rheumatism and inflammation (Goyal et al., 2003). Onion (*Allium cepa* L.) is used for insect bite, relieves pain and helps reduce irritation, as a repellent against insects.

Table 11. Plants identified from the family *Liliaceae* with their uses and local name.

Species	Local name	Parts used	Aliments	References
Asparagus racemosus Willd.	Shotomul	Leaves, root	Diabetes, heart diseases	Rahman *et. al.*, 2008; Rahmatullah *et. al.*, 2009; Tewari *et. al.*, 2000; Rahmatullah *et. al.*, 2010c;

Species	Local name	Parts used	Aliments	References
				Rahmatullah et. al., 2010d
Crinum asiaticum L.	Go-roshun	Leaves	Dysentery in cattle	Rahmatullah et. al., 2009; Ling et. al., 2009; Rahmatullah et. al., 2010b; Rahmatullah et. al., 2010d; Chowdhury et. al., 2009

Umbelliferae

The Umbelliferae or carrot family consists of plants usually produce an essential oil, and many of the species has medicinal value. Only one species in Bangladesh found to used for traditional medicine. Thankunis (*Centella asiatica*) is one of ancient herbal remedy which is reputed to restore youth, memory and longevity (Kapoor, 1990), to retard age and prevent dementia, combined with milk is given to improve memory (Manyam1999), used to treat rheumatic disorders, which suggests it may have anti-inflammatory effects, used for combating physical and mental exhaustion (Brinkhaus et al., 2000; Duke and Ayensu, 1985).

Table 12. Plants identified from the family Umbelliferae with their uses and local name.

Species	Local name	Parts used	Aliments	References
Centella asiatica (L.) Urb.	Thankuni, Adoni	Whole plant, Leaves	Blood purifier, fever, diabetes. Cataract, Gastrointestinal disorders	Rahmatullah *et. al.*, 2010b; Uddin *et. al.*, 2006; Ling *et. al.*, 2009; Melanie-Jayne *et. al.*, 2003; Ariful *et. al.*, 2010; Rahmatullah *et. al.*, 2010c; Rahman a *et. al.*, 2010; Mukul *et. al.*, 2007; Chowdhury *et. al.*, 2009; Khan & Rashid, 2006

Araceae

The Araceae, comprising 3200 species, is mainly tropical and is distributed worldwide (Croat 1979, 1994). In Bangladesh eight species are reported to be used for medicinal purpose and mainly including different types of kochu gach. Tuber of kata-kochu (*Lasia spinosa* L.) used to treatment edema, piles, constipation (Rahmatullah et. al., 2009; Rahmatullah et. al., 2010e), stem of bish kochu (*Colocasia nymphaeifolia* Vent.) used to treated coughs in children (Rahman a et. al., 2010),. leaves, stem of ghatkol (*Typhonium trilobatum* L.) for blood dysentery (Rahmatullah *et. al.*, 2010d), lower part of plat and stem of maan-kochu (*Alocasia macrorrhizos*) used to treated pus in ears, decreased eyesight, swelling of throat in cattle (Rahmatullah *et. al.*, 2010c, Rahmatullah *et. al.*, 2009). Root of *Acorus calamus* used in ayurvedic medicine as promoting rasayana effects (Manyam, 1999) and has been used to treat memory loss, also shown antioxidant activity in vitro (Acuna et al., 2002).

Table 13. Plants identified from the family Araceae with their uses and local name.

Species	Local name	Parts used	Aliments	References
Lasia spinosa (L.) Thwaites	Kata-kochu	Tuber	Edema, piles, constipation.	Rahmatullah *et. al.*, 2009; Rahmatullah *et. al.*, 2010e
Amorphophallus Campanulatus Blume ex Decne	Horong	Stem	Sudden paralysis.	Rahman a *et. al.*, 2010
Colocasia nymphaeifolia Vent.	Bish kochu	Stem	Coughs in children.	Rahman a *et. al.*, 2010
Typhonium trilobatum	Ghatkol	Leaves, stem	Blood dysentery.	Rahmatullah *et. al.*, 2010d

Medicinal plants in Bangladesh and its curative effects for disease

Species	Local name	Parts used	Aliments	References
(L.) Schott				
Alocasia macrorrhizos	Maan-kochu	Lower part of plant, stem	Pus in ears, decreased eyesight, swelling of throat in cattle.	Rahmatullah et. al., 2010c, Rahmatullah et. al., 2009
Colocasia esculenta (L.) Schott stem	Jongla-kochu	Whole plant, stem	Anti-hemorrhagic, blood purifier, to strengthen bones.	Rahmatullah et. al., 2009 ; Chowdhury et. al., 2009
Typhonium trilobatum (L.) Schott	Kharkon-shak	Leaves	Loss of appetite, mucus.	Rahmatullah et. al., 2009
Colocasia esculenta (L.) Schott	Bon-kochu	Whole plant	Severe jaundice, digestive aid, constipation.	Rahmatullah et. al., 2010c,

Convolvulaceae

About 1650 species present in Convolvulaceae family which are widely distributed in tropical, subtropical, and temperate regions, mostly herbaceous vines, but also trees, shrubs and herbs. In these study we found six species to used as medicinal plant in Bangladesh. Whole plant of kolmi-shak (*Ipomoea aquatic*) is used for the treatment of stomach problem (Rana *et. al.*, 2010; Ling et. al., 2009; Tewari et. al., 2000; Rahmatullah et, al, 2010b). Roots of vui-kumra (*Ipomoea mauritiana*) used for increased lactation (Hossan et. al., 2010; Rahmatullah et. al., 2010b; Rahmatullah et. al., 2010d). Leaves, stem and roots of vrommi lota (*Ipomoea mauritiana* Jacq.) used for excessive bile acid secretion, infrequent urination, to increase strength, to increase lactation in nursing mothers, pain in bones, gastric pain (Rahmatullah et. al., 2010c).

Table 14. Plants identified from the family Convolvulaceae with their uses and local name.

Species	Local name	Parts used	Aliments	References
Ipomoea aquatica	Kolmi-shak	Whole plant	Stomach Problem	Rana et. al., 2010; Ling et. al., 2009; Tewari et. al., 2000; Rahmatullah et, al, 2010b
Ipomoea mauritiana	Vui-kumra	Root	Increase lactation	Hossan et. al., 2010; Rahmatullah et. al., 2010b; Rahmatullah et. al., 2010d
Ipomoea fistulosa Mart. ex Choisy	Dhol-kolmi	Leaves	Swelling due to fracture	Rana et. al., 2010; Mollik et. al., 2010; Rahmatullah et. al., 2009
Ipomoea mauritiana Jacq.	Vrommi lota	Leaves, stem, root	Excessive bile acid	Rahmatullah et. al.,2010c

Medicinal plants in Bangladesh and its curative effects for disease

Species	Local name	Parts used	Aliments	References
			secretion, infrequent urination, to increase strength, to increase lactation in nursing mothers, pain in bones, gastric pain	
Ipomoea batatas (L.) Lam.	Mishti alu	Leaves and tubers	Stomach ache vitamin deficiency	Mollik *et. al.*, 2010
Ipomoea quamoclit Linn.		Aerial parts		Hasan *et. al.*, 2009a

Asteraceae

The Asteraceae is the largest family of vascular plants has more than 22,750 species world wide, but in Bangladesh 12 species found to be used as medicinal plants. Leaves of keshraj (*Eclipta alba* L.) used to stop bleeding and leaves of bhringoraaz (*Sida cordifolia* L.) for the gingivitis, tooth problems (Rahmutullah et. al., 2010c). Leaves of genda-phool (*Tagetes erecta* L.) used for anti-hemorrhagic, leaves, stem and seeds of bhanga-lota or assam lota (*Mikania cordata*) for the stomachache, helminthiasis, sprain, fracture, stop bleeding (Rahmatullah, 2009), roots of hagra (*Xanthium indicum* J. Koenig ex Roxb.) used for rheumatism (Chowdhury, 2010).

Table 15. Plants identified from the family Asteraceae with their uses and local name.

Species	Local name	Parts used	Aliments	References
Mikania cordata (Burm.f.) B. L. Robinson	Assam lota	Leaves	Cuts and wounds. Leaves. juice is applied	Rahmatullah et. al., 2009; Rahman a et. al., 2010; Mukul et. al., 2007
Eclipta prostrata L.	Kalinjiri	Whole plant	Jaundice	Chowdhury et. al., 2010
Mikania cordata (Burm.f.) B. L.	Riot lota, Pakistani lota	Leaves	Cuts and wounds, ulcer.	Rahmatullah et. al., 2010c; Chowdhury et. al., 2010
Spilanthes Paniculata Wall. ex DC.	Ot phol	Flower	Toothache, tooth infections.	Chowdhury et. al., 2010
Xanthium indicum J. Koenig ex Roxb.	Hagra	Root, Leaves	Rheumatism, Antidiarrhoeal Activity	Chowdhury et. al., 2010; Hasan et. al., 2009a; Hasan et. al., 2009b; Akter et. al.,

Medicinal plants in Bangladesh and its curative effects for disease

Species	Local name	Parts used	Aliments	References
				2009
Blumea lacera DC	Bon-mulo	Root	Helminthiasis, passing of semen in urine	Rahmatullah et. al., 2010d
Mikania cordata (Burm.f.)	Jarmany lota	Leaves, stem	Cooling effect	Rahmatullah et. al., 2010d
Spilanthes acmella (L.) Murray	Nak phool	Flower	Tooth ache	Rahmatullah et. al., 2010d
Mikania cordata (Burm.f.) B.L Robinson	Bhanga-lota	Stem, seed	Bloating, stomachache, helminthiasis, sprain, fracture	Rahmatullah et. al., 2009
Tagetes erecta L.	Genda-phool	Leaves	Anti-hemorrhagic	Rahmatullah et. al., 2010c; Rahmatullah et. al., 2009
Eclipta alba (L.) Hassk.	Keshraj	Leaves	Anti-hemorrhagic	Rahmatullah et. al., 2010c; Patil and Bhaskar, 2006
Sida cordifolia L.	Bhringoraaz	Leaves	Gingivitis, tooth problems	Rahmatullah et. al., 2010c
Xanthium strumarium,	Ghaghra shak	Dry leaves, Roots	Asthma Skin disease Dental pain Rheumatism Jaundice	

Bromeliaceae

Bromeliaceae is a family of monocot flowering plants of around 3,170 species worldwide of which only one species reported to be used in Bangladesh as medicinal plant. Pineapple (*Ananas comosus*) contains compounds that relieve inflammation and swelling *Ananas comosus*, used for ailments of the CNS, also administered for neurasthenia, melancholy, sadness and forweakness of memory (Wolters, 1994). The proteolytic enzyme bromelaine is the best known bioactive component from *Ananas* sp. is, which is responsible for the positive effects on digestion, used for rapid weight-loss.

Table 16. Plants identified from the family Bromeliaceae with their uses and local name.

Species	Local name	Parts used	Aliments	References
Annas comosus (L.) Merr.	Anarosh	Ripe fruit	Fever, intestinal worm	Mukul *et. al.*, 2007; Chowdhury *et. al.*, 2009

Others Families

Leaves and fruit of telakucha (*Coccinia grandis* L., Cucurbitaceae family) used for the treatment of typhoid disease, eczema, leukoderma, voigt lesions of the tongue, leukoderma (Rana *et. al.*, 2010; Rahmatullah *et. al.*,

2010b; Rahman et. al., 2008; Hossan et. al., 2010; Rahmutullah et. al., 2009; Rahmatullah et. al., 2010c; Chowdhury et. al., 2010; Mukul et. al., 2007; Chowdhury et. al., 2009; Rahmatullah et. al., 2010e). Leaves and fruit of aam (*Mangifera indica* L., Anacardiacea family) used for eye diseases, antidote to poison, edema, seed cholera, dysentery, diabetes. stomach ache, cuts and wounds (Hossan et. al., 2010; Ling et. al., 2009; Rahman a et. al., 2010; Rahmatullah et. al., 2010b; Rahmatullah et. al., 2010c; et. al., 2010; Rahmatullah et. al., 2010d; Rahman et. al.,2008; Chowdhury et. al., 2009; Rahmatullah et. al., 2010e). Whole plant of ghrit-kumari (*Aloe vera*, Aloaceae family) used for the treatment of laxative, appetizer, alopecia, asthma, tuberculosis (Rahman et. al.,2008; Hossan et. al.,2010; Ling et. al., 2009; Rahmatullah et. al., 2010b; Rahmatullah et. al., 2010e; Khan & Rashid, 2006). Fresh juice, seed coat and tender of narkel (*Cocos nucifera* Linn., Arecaceae family) used for hair falls, burns (Rana et. al., 2010; Rahman et. al., 2008; Rahmatullah et. al., 2010c; Hasan et. al., 2009a; Mukul et. al., 2007; Chowdhury et. al., 2009; Rahmatullah et. al., 2010e). Flower of joba (*Hibiscus rosasinensis* L., Malvaceae family) used to treatment of dysentery (Ling et. al., 2009; Mollik et. al., 2010; Rahmatullah et. al., 2010c; Rahmatullah et. al., 2009; Rahmatullah et. al., 2010d; Chowdhury et. al., 2009). Leaves of mendi (*Lawsonia inermis* L., Lythraceae family) used for dandruff and grey hair (Rana et. al.,

2010; Rahman et. al., 2008; Rahmatullah et. al., 2010c; Chowdhury et. al., 2009; Rahmatullah et. al., 2010d). Leaves and roots of Nishinda *(Vitex negundo*, Lygodiaceae family) used for the treatment of cold, cough, asthma, fever and diuretic (Rahman et. al., 2008; Mollik et. al., 2010; Rahmatullah et. al., 2010c; Rahmatullah et. al., 2009; Chowdhury et. al., 2009; Khan & Rashid, 2006). Whole plant of Durba (*Cynodon dactylon* L., Poaceae family) used for coughs, cuts and wounds treatment (Rana et. al., 2010; Rahman et. al., 2010; Uddin et. al., 2006; Mollik et. al., 2010;Rahmatullah et. al., 2010b; Rahmatullah et. al., 2010c; Rahmatullah et. al., 2009; Chowdhury et. al., 2010). Dandokolos (*Leucas aspera* LINK, *Labiatae* family) is a aromatic herb in Bangladesh, the whole plant is used for analgesic-antipyretic, antirheumatic,antiinflammatory, and antibacterial treatment, etc., and its paste is applied topically to inflamed areas (Ahmed et al., 2010; Rahman et. al., 2007; Saha et. al., 2008). The fruits and leaves of sajna (*Moringa oleifera, Moringaceae*family) are used in hepatitis and spleen inflammation, arthritis, tetanus and paralysis. The leaves possess antihypertensive activity (Ghani, 2000). The leaves of sajna also used as a vegetable.

Medicinal plants in Bangladesh and its curative effects for disease

Table 17. Plants identified from other families with their uses and local name.

Family name	Scientific name	Local name	Parts Used	Ailments	Reference
Acanthaceae	*Acanthus ilicifolius* L	Fereng-jubang R	Roots	Leucorrhoea	Hossan et. al., 2010
	Hygrophila spinosa T. Anders	Kuel-kharha	Whole plant	Leucorrhoea	Hossan et. al., 2010
	Andrographis paniculata	Kalomegh	Whole plant, Leaves	Worm infection, fever, dysentery, diarrhoea and tonic.	Rahman et. al., 2008; Rahmatullah et.al, 2010c; Uddin et. al., 2006; Tewari et. al., 2000; Khan & Rashid, 2006
	Hemigraphis hirta (Vahl)T. Anderson	Buripal	Leaves	Headache, passing of semen in urine.	Rahmatullah et. al., 2010d
	Justicia adhatoda L.	Bashok	Leaves, bark	Leucorrhea, chronic respiratory disorders, Fever, cold, coughs.	Rana et. al., 2010, Rahman et. al., 2008; Rahman a et. al., 2010; Rahmatullah

45

Family name	Scientific name	Local name	Parts Used	Ailments	Reference
					et. al., 2010b; Rahmatullah et. al., 2010c; Rahmatullah et. al., 2009; Chowdhury et. al., 2010 ; Mukul et. al., 2007; Chowdhury et. al., 2009; Khan & Rashid, 2006
Adiantaceae	Acrostichum aureum L.	Mouchaipang	Leaves	Leucorrhoea	Hossan et. al., 2010
Aizoaceae	Tetragonia tetragonoides (Pall.) Kuntze	Shonta	Whole plant	eye diseases	Hossan et. al., 2010; Rahmatullah et. al., 2010e
Aloaceae	Aloe vera (L.) Burm.f.	Ghritkumari	Whole plant	Laxative, appetizer, alopecia, asthma, tuberculosis	Rahman et. al.,2008; Hossan et. al.,2010; Ling et. al., 2009; Rahmatullah et. al., 2010b; Rahmatullah et. al., 2010e; Khan & Rashid, 2006
Anacardiacea	Mangifera indica L	Aam	Leaves, fruit	Eye diseases,	Hossan et. al., 2010;

Medicinal plants in Bangladesh and its curative effects for disease

Family name	Scientific name	Local name	Parts Used	Ailments	Reference
				antidote to poison, edema, seed cholera, dysentery, diabetes. Stomach ache, cuts and wounds,	Ling et. al., 2009; Rahman a et. al., 2010; Rahmatullah et. al., 2010b; Rahmatullah et. al., 2010c; et. al., 2010; Rahmatullah et. al., 2010d; Rahman et. al., 2008; Chowdhury et. al., 2009; Rahmatullah et. al., 2010e
	Mangifera longipes Griffith	Uri aam	Bark	Loss of sensitivity in skin, chronic mucus, continuous Griffith sneezing,	Chowdhury et. al., 2010
	Spondias pinnata (J.G. Konig ex L. f.) Kurz.	Amra	Bark	Dysentery with white mucus.	Rahmatullah et. al., 2010d
Arecaceae	Borassus flabellifer L.	Tal	Root, fruit	Cancer, edema, epilepsy, boil.	Rahmatullah et. al., 2010e
	Cocos nucifera	Narkel	Fresh	Hair falls,	Rana et. al.,

Family name	Scientific name	Local name	Parts Used	Ailments	Reference
	Linn.		juice, Tender, seed coat	burns	2010; Rahman et. al., 2008; Rahmatullah et. al., 2010c; Hasan et. al., 2009a; Mukul et. al., 2007; Chowdhury et. al., 2009; Rahmatullah et. al., 2010e
Cucurbitaceae	Momordica charantia L.	Korla	Root, seed, fruit	Cancer, rheumatoid arthritis, helminthiasis. Diabetes	Rana et. al., 2010; Rahmatullah et. al., 2010; Rahmatullah et. al., 2009; Rahmatullah et. al., 2010d; Chowdhury et. al., 2009 ; Rahmatullah et. al., 2010e
	Coccinia grandis L.	Telakucha	Leaves, fruit	Typhoid disease, eczema, leukoderma, Voigt lesions of the tongue, leukoderma	Rana et. al., 2010; Rahmatullah et. al., 2010b; Rahman et. al., 2008; Hossan et. al., 2010; Rahmutullah et. al., 2009; Rahmatullah et. al.,

Medicinal plants in Bangladesh and its curative effects for disease

Family name	Scientific name	Local name	Parts Used	Ailments	Reference
					2010c; Chowdhury et. al., 2010; Mukul et. al., 2007; Chowdhury et. al., 2009; Rahmatullah et. al., 2010e
	Momordica cochinchinensis (Lour.) Spreng.	Boro kakroal, Karol gach	Seed, root	Rheumatism (seed), pain, abscess (root)	Rahmatullah et. al., 2009; Rahmatullah et. al., 2010d
Phyllanthaceae	*Phyllanthus reticulatus* Poir.	Chitki	Leaves, bark	Edema, constipation, cooling	Rahmatullah et. al., 2010e
	Baccaurea ramiflora Lour.		Fruit pericarp		Hasan et.al., 2009a
	Feronia limolia Linn.		Leaves		Hasan et. al., 2009a
Caesalpiniaceae	*Cassia alata* L	Daad	Leaves	Scabies, skin diseases eczema, ringworm	Rahman et. al., 2008; Chowdhury et. al., 2009; Khan & Rashid, 2006
Puniacaceae	*Punica granatum* L.	Dalim	Leaves	Worm in Intestine, Blood dysentery	Rahman et. al., 2008; Ling et. al., 2009; Rahmatullah et. al.,

Family name	Scientific name	Local name	Parts Used	Ailments	Reference
					2010d; Chowdhury et. al., 2009; Hasan et. al., 2009a;
Plantaginaceae	*Plantago ovate* Forst.	Isopgul	Seed, coat, Seeds	Heat stroke, gastric pain, constipation, Sexual weakness in males	Tewari et. al., 2000; Chowdhury et. al., 2009
Malvaceae	*Hibiscus rosasinensis* L.	Joba	Flower	Dysentery	Ling et. al., 2009; Mollik et. al., 2010; Rahmatullah et. al., 2010c; Rahmatullah et. al., 2009; Rahmatullah et. al., 2010d; Chowdhury et. al., 2009
	Urena lobata L.	Adira kata	Root	Urinating during sleep	Rahman a et. al., 2010
	Hibiscus mutabilis L.	Cotton Rose, Chinese Rose	Leaves, flower	Fistulae, pustules tumours, diuretic and bacterial infection	Ling et. al., 2009

Medicinal plants in Bangladesh and its curative effects for disease

Family name	Scientific name	Local name	Parts Used	Ailments	Reference
	Sida cordifolia L.	Mithe berela	Leaves, root	Dysentery (presence of mucus in stool)	Rahmatullah et. al., 2010b; Rahmatullah et. al., 2010c; Rahmatullah et. al., 2010d; Ahmed et.al., 2010
Musaceae	*Musa spp* Var.	Kola	Green fruit	Dysentery, helminthiasis	Rahmatullah et. al., 2009; Chowdhury et. al., 2009
	Musa paradisiaca L.	Bichi kola	Root	Dysentery. Roots of *Ficus hispida* (1 chatak,	Rahmatullah et. al., 2010
	Entese superbum (Roxb) cheesm	Jengli keli	Seeds	Dog bite	Patil and Bhaskar, 2006
Lythraceae	*Lawsonia inermis* L.	Mendi	Leaves	Dandruff and grey hair	Rana et. al., 2010; Rahman et. al., 2008; Rahmatullah et. al., 2010c; Chowdhury et. al., 2009; Rahmatullah et. al., 2010d
Annonaceae	*Polyalthia longifolia*	Debdaru	Bark	Dysentery, itch	Rahmatullah et. al., 2010b

Family name	Scientific name	Local name	Parts Used	Ailments	Reference
Annonaceae	*Annona squamosa* L.	Ata	Leaf, fruit, seed	Abortifacient, helminthiasis, colic, itch.	Rahmatullah *et. al.*, 2010e
Amaryllidaceae	*Curculigo orchioides*	Taalmuli	Root	Blood coming out through anus Root Gaertn.	Rahmatullah *et. al.*, 2010b
Aristolochiaceae	*Aristolochia indica* L	Ishwarmul	Leaves, root	Biliary pain, infection	Khan *et. al.*, 2006; Rahmatullah *et. al.*, 2010b; Rahmatullah *et. al.*, 2010d; Khan & Rashid, 2006
Cactaceae	*Cereus grandiflorus*(L.) P.Mill.	Kaat-raaz	Inner portion	Acidity	Rahmatullah *et. al.*, 2010b
	Opuntia dillenii (Ker-Gawl.) Haw.	Foni-monsha	Leaves	Headache, menstrual pain, pain in leg, passing of	Rahmatullah *et. al.*, 2010d

Medicinal plants in Bangladesh and its curative effects for disease

Family name	Scientific name	Local name	Parts Used	Ailments	Reference
				semen in urine	
Piperaceae	*Piper betel* L	Paan	Leaves	Flatulence, indigestion Congestion in muscles cuts and wounds	Rana *et. al.*, 2010; Ling *et. al.*, 2009 Rahmatullah *et. al.*, 2010c, Rahmatullah *et. al.*, 2009
	Piper longum L	Pipol	Leaves	Indigestion, stomach disorders	Rahman a *et. al.*, 2010; Rahmatullah *et. al.*, 2010b; Rahmutullah *et. al.*, 2009
	Piper chaba W. Hunter	Chui gach	Bark	Coughs	Rahmatullah *et. al.*, 2010d
Clusiaceae	*Mesua nagassarium* (Burm.f.) Kosterm.	Nageshwar	Leaves, bark, Flower	Inability to bear children in women, Constipation	Rahmatullah *et. al.*, 2009;
	Garcinia cowa Roxb.	Kau-pata	Leaves	Gastrointestinal problems, diarrhea	Rahmatullah *et. al.*, 2010c
Crassulaceae	*Kalanchoe pinnata* (Lam.) Pers	Pathor kuchi	Whole plant	Gall bladder stones, Pain from	Rana *et. al.*,2010; Rahman *et.*

Family name	Scientific name	Local name	Parts Used	Ailments	Reference
				piles, Stomach pain	al.,2008; Rahmatullah et. al.,2010b; Rahmatullah et. al.,2010d; Chowdhury et. al., 2009
	Kalanchoe pinnata (Lam.) Pers	Thanda pata	Leaves	Kidney stones	Rahman et. al.,2010
Ebenaceae	*Diospyros peregrine* (Gaertn.) Gürke	Gab gach	Whole plant	Leucorrhea, thorn-induced infections, gangrene, cough, mucus, biliary diseases, blood purifier	Rahmatullah et. al.,2010b; Rahmatullah et. al.,2010c; Rahmatullah et. al.,2009; Chowdhury et. al.,2010; Rahmatullah et. al.,2010d
Menispermaceae	*Tinospora cordifolia* (Willd.) Hook.f.	Guloncho lota	Whole plant	Measles, burning sensations, cough, mucus, fever, helminthiasis,	Rahmutullah et. al.,2010b; Hasan et. al., 2009a
Pandanaceae	*Pandanus odoratissimus* L.f.	Keya	Leaves	Obesity	Rana et. al.,2010; Rahmatullah et. al.,2010b
Rubiaceae	*Paederia foetida* L.	Gondho-vadule	Whole plant	Cough, mucus, Rheumati	Mollik et. al.,2010;

Medicinal plants in Bangladesh and its curative effects for disease

Family name	Scientific name	Local name	Parts Used	Ailments	Reference
				sm, Pain, dysentery, Indigestion, Stomach pain	Rahmatullah et. al.,2010c; Rahmatullah et. al.,2010d
	Anthocephalus chinensis (Lam.) A. Rich. ex Walp.	Kodom gach	Young leaves, stems	Headache	Rahman et. al.,2008; Mollik et. al.,2010; Rahmatullah et. al.,2010b; Rahmatullah et. al.,2009; Chowdhury et. al.,2010; M Rahmatullah et. al., 2010d
Amaranthaceae	Achyranthes aspera L	Withlenka	Whole plant	To increase libido. The whole plant is cut into small pieces and tied to the waist.	Mollik et. al., 2010; Rahman et. al., 2010; Rahmatullah et. al., 2010c
	Alternanthera sessilis (L.) R. Br. ex DC.	Sachi	Whole plant, Leaves	Dysentery	Mollik et. al., 2010; Rahman et. al., 2010

Family name	Scientific name	Local name	Parts Used	Ailments	Reference
	Achyranthes aspera	Apang	Root, bark, Leaves	Sciatica, abortion, eczema and wound	Rahman *et. al.*, 2008; Mollik *et. al.*, 2010; Rahman *et. al.*, 2010
	Amaranthus spinosus L	Kanta bujri, Khuira kanta	Leaves, stem, root	Rheumatic pain, Bone pain, blood or pus in urine	Rahmatullah *et. al.*, 2009; Mollik *et. al.*, 2010; Chowdhury *et. al.*, 2010
	Achyranthes aspera L. var. *nigro-olivacea* Suess.	Chirchiri	Roots	Leucorrhoea, menstrual problems	Hossan *et. al.*, 2010
Sterculiaceae	*Abroma augusta* L.f.	Ulot kombol	Whole plant	Passing of semen with urine, Astringent.	Rahmatullah *et. al.*, 2010b; Hossan *et. al.*, 2010; Rahmatullah *et. al.*, 2010c; Rahmatullah *et. al.*, 2010d; Mukul *et. al.*, 2007

Medicinal plants in Bangladesh and its curative effects for disease

Family name	Scientific name	Local name	Parts Used	Ailments	Reference
Compositae	*Kesuti Eclipta alba*	Kesuti	Whole plant	Spleen disorder, stomach disorder, dermatitis, constipation, headache	Rahmatullah, 2010
Leguminosae	*Arachis hypogaea*	Cheena badam	kernel of seed.	diabetes, gonorrhea, tonic, wounds.	Rahmatullah, 2010
Cruciferae	*Brassica napus*	Sharisa	leaves, seeds.	Fever, common cold, stomachache, itching, headache	Rahmatullah, 2010
Bombacaceae	*Bombax ceiba* L.	Shimul tula	Leaves, root, bark	Stoppage of urination and defecation	Rahmatullah *et. al.*, 2010b; Ling *et. al.*, 2009; Rana *et. al.*, 2010; Mollik *et. al.*, 2010; Chowdhury *et. al.*, 2010; Rahmatullah *et. al.*, 2010d; Patil and Bhaskar, 2006
	Bombax cieba	Shimul	roots, leaves, barks	Fever, leprosy, pox, diarrhea,	Mollik *et. al.*, 2010; Rahman *et. al.*, 2010;

Family name	Scientific name	Local name	Parts Used	Ailments	Reference
				dysentery, weakness, impotency, gastric ulcer.	Rahmatullah et. al., 2010b; Rahmatullah et. al., 2010c
	Heliotropium indicum L	Hatirshur	Whole plant	Frequent urination	Mollik et. al., 2010; Rahman et. al., 2010; Chowdhury et. al., 2010
Costaceae	Costus speciosus (J. König.) Sm	Kaeu	Root	Dysentery, stomach ache Itch, scabies	Mollik et. al., 2010; Rahman et. al., 2010; Rahmatullah et. al., 2009
Cyperaceae	Cyperus rotundus L.	Kellar bori	Seed	Helminthiasis, frequent urination	Mollik et. al., 2010; Rahman et. al., 2010
	Cyperus kyllingia Endl.	Badlar shuta	Root	Aphrodisiac, urinary problems	Rahmatullah et. al., 2010d
Lauraceae	Litsea glutinosa (Lour.). C.D.Robins	Bijol	New Leaves	Low semen density	Rahman et. al., 2010
	Cinnamomum camphora (L.) J. Presl	Korpur	Leaves, bark	Skin diseases and toothache	Rahmatullah et. al., 2009
	Cinnamomum	Daruc	Bark	Debility, dysentery	Rahmatullah

Medicinal plants in Bangladesh and its curative effects for disease

Family name	Scientific name	Local name	Parts Used	Ailments	Reference
	zeylanicum Blume	hini		, growth retardation	et. al., 2009
	Cinnamomum tamala (Buch.-Ham.) Nees & Eberm.	Tejpata	Leaves	Coughs, common colds	Ling et. al., 2009; Rahmatullah et. al., 2010c; Rahmatullah et. al., 2009; Khan & Rashid, 2006
Lygodiaceae	Lygodium flexuosum	Dheki-shak	Leaves	Diarrhea, stomachache	Mollik et. al., 2010; Rahmatullah et. al., 2009
	Lygodium flexuosum (L.) Sw	Kukur mutha Kukur shuka Bon tamak	Root	Dysentery	Mollik et. al., 2010; Rahman et. al., 2010; Rahmatullah et. al., 2009
	Vitex negundo	Nishinda	Leaves, root	Cold, cough, asthma, fever and diuretic.	Rahman et. al., 2008; Mollik et. al., 2010; Rahmatullah et. al., 2010c; Rahmatullah et. al., 2009; Chowdhury et. al., 2009; Khan & Rashid, 2006

Family name	Scientific name	Local name	Parts Used	Ailments	Reference
	Lippia alba (Mill.) N.E. Br. ex Britton & P. Wilson	Khuria	Leaves	Cuts and wounds	Rahman *et. al.*, 2010
	Nyctanthes arbor tristis L.	Shefali phool	Leaves, flower	Fever, Cough	Rahmatullah *et. al.*, 2010c; Rahmatullah *et. al.*, 2010b; Rahmatullah *et. al.*, 2010d; Chowdhury *et. al.*, 2010
	Phyla nodiflora (L.) Greene	Saitta okra, Okra pata	Leaves, stem, bark	Pain, back or waist pain due to rheumatism	Rahmutallah, 2010
	Clerodendrum viscosum Vent.	Bhat	Leaves, stem	Fever in children, toothache, pain in gums	Rahmatullah *et. al.*, 2009; Chowdhury *et. al.*, 2010; Rahmatullah *et. al.*, 2010d; Chowdhury *et. al.*, 2009
Menispermaceae	*Stephania japonica (Thunb.) Miers*	Takpata	Leaves	Low sperm count, pain, dysentery	Mollik *et. al.*, 2010; Rahman *et. al.*, 2010

Medicinal plants in Bangladesh and its curative effects for disease

Family name	Scientific name	Local name	Parts Used	Ailments	Reference
	Tinospora crispa (L.) Hook.f. & Thoms.	Guntai	Whole plant	Low sperm count	Rahman *et. al.*, 2010
	Stephania japonica (Thunb.)	Doi-pata	Leaves	Helminthiasis, skin diseases	Mollik *et. al.*, 2010; Rahmatullah *et. al.*, 2009
Myrtaceae	*Syzygium cumin*	Kalo jaam	Leaves	Low sperm count	Rahman *et. al.*, 2008; Mollik *et. al.*, 2010; Rahman *et. al.*, 2010; Rahmatullah *et. al.*, 2010c
	Syzygium cumini (L.) Skeels	Jaam	Seed, bark	Diabetes	Rahman *et. al.*, 2008; Mollik *et. al.*, 2010; Rahmatullah *et. al.*, 2010d; Hasan *et. al.*, 2009a; Chowdhury *et. al.*, 2009
	Syzygium Alaccense (L.) Merr. & L. M. Perry	Jamrul gach	Leaves	Helminthiasis, stomach ache, Cold, itch, waist pain	Mollik *et. al.*, 2010; Rahmatullah *et. al.*, 2010c
	Psidium guajava L.	Piyara gach	Leaves	Gastric problems, cuts and wounds	Rahman *et. al.*, 2008; Rahmatullah *et. al.*, 2009;

Family name	Scientific name	Local name	Parts Used	Ailments	Reference
					Rahmatullah et. al., 2010d
Poaceae	*Cynodon dactylon* (L.) Pers	Durba	Whole plant	Coughs, cuts and wounds	Rana et. al., 2010; Rahman et. al., 2010; Uddin et. al., 2006; Mollik et. al., 2010; Rahmatullah et. al., 2010b; Rahmatullah et. al., 2010c; Rahmatullah et. al., 2009; Chowdhury et. al., 2010
	Eleusine sp.	Gora durba	Whole plant	Infections	Rana et. al., 2010; Rahman et. al., 2010
Scrophulariaceae	*Scoparia dulcis* L	Bon-dhonya	Leaves	Hechki (local term for continuous hiccups)	Rahman et. al., 2010; Mollik et. al., 2010
	Scoparia dulcis L	Chinigura	Leaves	Diabetes	Mollik et. al., 2010; Rahmatullah et. al., 2009
Smilacaceae	*Smilax zeylanica* L	Bagha-chora, Kumai ra lota	New Leaves	Leukorrhea	Rahman et. al., 2010; Rahmatullah et. al., 2010b
Urticaceae	*Laportea crenulata*	Damma	Root	Skin disorders, infections	Rahman et. al., 2010

Medicinal plants in Bangladesh and its curative effects for disease

Family name	Scientific name	Local name	Parts Used	Ailments	Reference
	Gaudich.-Beaup				
	Boehmeria macrophylla	Chulka ni-pata	Leaves	Poisonous insect bite	Rahmatullah et. al., 2009
Asclepiadaceae	*Calotropis procera* (Ait.) Ait.f.	Akondo	Leaves, sap	Dog bite, Fractures, asthma and wounds	Rana et. al., 2010; Rahman et. al., 2008; Rahmatullah et. al., 2009; Rahman et. al., 2010; Chowdhury et. al., 2010; Rahmatullah et. al., 2010d
Cuscutaceae	*Cuscuta reflexa* Roxb.	Shorno lota	Wholeplant	Rheumatic fever, Lesion, Jaundice	Rahmatullah et. al., 2010c; Uddin et. al., 2006
	Cuscuta reflexa Roxb	Shunnyo-lota	Stem	Blood clotting, jaundice	Rahmatullah et. al., 2009; Chowdhury et. al., 2010
Polygonaceae	*Polygonum hydropiper* L.	Bish katali	Whole plant	Pain, swellings, fractures	Mollik et. al., 2010; Rahmatullah et. al., 2010c; Rahmatullah et. al., 2009; Chowdhury et. al., 2010

Family name	Scientific name	Local name	Parts Used	Ailments	Reference
Scrophulariaceae	*Scoparia dulcis* L.	Mishri dana, Chini champa	Leaves	Dysentery in children	Mollik *et. al.*, 2010; Chowdhury *et. al.*, 2010
Agavaceae	*Sansevieria trifasciata* Prain	Shaper gach	Whole plant, Leaves	Snake repellent	Mollik *et. al.*, 2010
Aristolochiaceae	*Aristolochia indica* L.	Iche gach	Root	Snake bite	Hossan *et. al.*, 2010
Chenopodiaceae	*Chenopodium ambrosioides* L.	Agnish war	Leaves	Acidity, burns	Rahmatullah *et. al.*, 2010d;;
Commelinaceae	*Commelina benghalensis* Linn.	Kengra gach	Leaves, Aerial parts	Pain	Rahmatullah *et. al.*, 2010d; Hasan *et. al.*, 2009a; Hasan *et. al.*, 2010
Flacourtiaceae	*Flacourtia indica* (Burm.f.) Merr.	Bujir gach, Kata gach	Leaves	Stomach ache	Rahmatullah *et. al.*, 2009; Rahmatullah *et. al.*, 2010d
Lecythidaceae	*Barringtonia racemosa* (L.) Roxb.	Moha shomudro gach	Leaves	Snake bite, snake repellent	Rahmatullah *et. al.*, 2009; Rahmatullah *et. al.*, 2010

Medicinal plants in Bangladesh and its curative effects for disease

Family name	Scientific name	Local name	Parts Used	Ailments	Reference
Oxalidaceae	*Oxalis lobata* Sims	Shushni-pata	Whole plant	Stomache, debility	Rahmatullah *et. al.*, 2009
Cappardaceae	*Crataeva religiosa* G. Forst.	Boinna	Leaves	Tumor	Rahmatullah *et. al.*, 2009
Moringaceae	*Moringa oleifera* Lam.	Sajna gach	Skin of fruit	Rheumatism, ear disease, headache	Rahman *et. al.*, 2008; Bhowmik *et. al.*, 2008; Rahmutullah *et. al.*, 2009; Rahmutullah *et. al.*, 2010a; Rahmutullah *et. al.*, 2010c; Rahmatullah *et. al.*, 2010d
Rhamnaceae	*Ziziphus jujuba* Mill.	Kul boroi	Leaves	Voice hoarseness	Rahman *et. al.*, 2008; Rahmatullah *et. al.*, 2010d; Hossan *et.al.*, 2010
Rosaceae	*Rosa damascena* Mill.	Golap	Flower	Appetite stimulant	Rahmatullah *et. al.*, 2010d ; Rahmatullah *et. al.*, 2010de;

Family name	Scientific name	Local name	Parts Used	Ailments	Reference
Sapotaceae	*Manilkara zapota* (L.) P. Royen	Chobeda gach	Fruit	Vitamin supplementation	Rahmutullah *et. al.*, 2009; Rahmatullah *et. al.*, 2010d
Cannabaceae	*Trema orientalis* (L.)	Kath-gach	Leaves	Dysentery, tiredness due to heat	Rahmutullah *et. al.*, 2009
Asphodelaceae	*Aloe barbadensis* Mill	Ghrito-kanchon	Leaves pulp	To keep head cool, dysentery	Rahmutullah *et. al.*, 2009
Apiaceae	*Centella asiatica* (L.) Urb.	Taka-pata	Leaves	Diarrhea, gastric problems	Mollik *et. al.*, 2010; Rahmutullah *et. al.*, 2009
	Centella asiatica (L.) Urban	Goal pata	Roots	Leucorrhoea, menstrual problems	Hossan *et. al.*, 2010; Mollik *et. al.*, 2010
Palmae	*Areca catechu* L	Betel Nut Palm, Areca Nut, Pinang	Inner seed	wounds, swellings and other skin afflictions	Ling *et. al.*, 2009
Leguminosae	*Bauhinia purpurea* L	Butterfly Tree	Root, leaves, Flower	common fever. cough treatment	Ling *et. al.*, 2009

Medicinal plants in Bangladesh and its curative effects for disease

Family name	Scientific name	Local name	Parts Used	Ailments	Reference
Leguminosae	*Dolichos lablab* L	Sim	Fruit, leaves, Seeds	Alcoholic intoxication, cholera, diarrhoea, gonorrhoea, leucorrhoea, nausea. cholera, rheumatism and sunstroke	Ling *et. al.*, 2009
Aspleniaceae	*Asplenium nidus* L	Bird's Nest Fern	Leaves	To ease labour pains	Ling *et. al.*, 2009
Balsaminaceae	*Impatiens balsamina* L	Dopati phool	Leaves and Flower	Eeczema, itches and insect bites, warts, cancer treatment and expectorant	Ling *et. al.*, 2009
Nyctaginaceae	*Mirabilis jalapa* L.	Four O'Clock Flowe, Shondha Moni phool	Leaves, flower, seed	Abscesses, blisters, and to relieve urticaria	Ling *et. al.*, 2009

Family name	Scientific name	Local name	Parts Used	Ailments	Reference
Papilionaceae	*Butea monosperma* (Lam.) Taub.		Leaves		Hasan *et. al.*, 2009a
Ceasalpiniaceae	*Caesalpinia pulcherrima* Linn.		Leaves		Hasan *et. al.*, 2009a
Dipterocarpaceae	*Hopea odorata* Roxb.		Leaves		Hasan *et. al.*, 2009a
Magnoliaceae	*Michelia champaca* Linn.		Leaves		Hasan *et. al.*, 2009a
Elaeocarpaceae	*Eleocarpus robustus* Roxb.	Jolpai	Fruit	Apathy to food	Chowdhury *et. al.*, 2009
Sapindaceae	*Swertia chirata* Ham.	Chirota	Whole plant	Gastric pain, diabetes, liver dis., fever	Chowdhury *et. al.*, 2009
Oxalidaceae	*Oxalis lobata* Sims	Amrul	Whole plant	Dysentery, diarrhea, coughs, stimulant.	Rahmatullah *et. al.*, 2010e

Medicinal plants in Bangladesh and its curative effects for disease

Family name	Scientific name	Local name	Parts Used	Ailments	Reference
Sonneratiaceae	*Sonneratia caseolaris* Linn.	Choila	Juices	astringent and antiseptic, arresting hemorrhage.	Ahmed et.al.,2010
Asclepiadaceae	*Asclepias parasitica* Wallich ex Hornemann	Bayupriya, Porgacha)	Aerial part	rheumatism.	Ahmed et.al.,2010
Labiatae	*Leucas aspera* LINK	Dando kolos	whole plant	analgesic-antipyretic, antirheumatic, antiinflammatory, and antibacterial treatment	Ahmed et al., 2010; Rahman et. al., 2007; Saha et. al., 2008

Khaton and Shaik

3. CHAPTER THREE: MEDICINAL PLANTS AND DISEASE

For searching of bioactive natural products targeting cancer related signaling pathways such as tumor necrosis factor (TNF)-related apoptosis-inducing ligand (TRAIL), it has been needed the intensive research for the medicinal plants Bangladesh (Ishibashi and Ohtsuki 2008, Ishibashi and Arai 2009). Hossain et al, 2003 reported that about 58 different medicinal plants were being used in rural Bangladesh to treat different diseases. The medicinal plants include ghritakumari, soto muli, arshagandha, thankuni, kalomegh, ashugandha, telakucha, har-bhangar gach, talamuli, ulatkambal, lazzaboti, different types of Chandal (like guruchandal, bhaichandal, raktachandal, turukchandal), arjun, pantharkuchi, sarpagandha, sonkhamul, ishwarmul, anantamul,, kalkashinda, vimraj, sishmoni, tulshi, tisi etc. These are used in various lesion, antacid, itching, for dysentery, skin disease, cold/cough, headache, fever, appetizer, paralysis, burning during urination and burning of palm and foot soles, mouth wash, dental plaque, rheumatic pain, pain killer, conjunctivitis, vomiting, fairness, blindness, bronchitis etc.

Cognitive enhancement

There have been numerous studies regarding the

cognitive enhancing activities of *Withania somnifera*. Withanoside IV or VI produced dendritic outgrowth in normal cortical neurons of isolated rat cells, whereas axonal outgrowth was observed in the treatment with withanolide A in normal cortical neurons (Tohda et al., 2005). Neuritic regeneration or synaptic reconstruction was induced by withanolide A, withanoside IV and VI in amyloid- (25–35)-induced damaged cortical neurons. In addition, these components also facilitated the reconstruction of post-synaptic and pre-synaptic regions in neurons, where severe synaptic loss had already occurred. *W. somnifera* extract, containing the steroidal substances sitoinodosides VII–X and withaferin A augmented learning acquisition and memory in both young and old rats (Ghosal et al., 1989). It enhanced AChE activity in the lateral septum and globus pallidus and decreased it in the vertical diagonal band. Receptor binding on the muscarinicM1 receptor was enhanced in the lateral and medium septum and in the frontal cortices. M2 receptor binding increased in cortical regions but did neither affect-aminobutyric acid (GABAA), benzodiazepine, nor NMDA receptor binding. The extract reversed ibotenic acid induced cognitive deficit and reversed the reduction in cholinergic markers, such as acetylcholine (Schliebs et al., 1997).

Rheumatoid arthritis

Rheumatoid arthritis is an autoimmune, chronic, systemic inflammatory disorder principally attacking the synovial joints (these joints achieve movement at the point of contact of the articulating bones). It is an autoimmune disorder, which cause inflammation of the joints and can cause inflammation of the tissue surrounding the joints Rheumatoid arthritis is quite common in Bangladesh. Methotrexate, a drug of choice in Bangladesh to treat rheumatoid arthritis, has been reported to give adverse effects in 27 out of the 38 patients studied (Ali, 1997).

The use of medicinal plants for treatment of rheumatic disorders is not new but is also practiced in the traditional medicinal systems of other countries of the world. A notable success story is that of the plant, *Harpagophytum procumbens* (Burch.) DC. ex Meisn. [Genus: Harpagophytum, Family: Pedaliaceae, English name: Devil's Claw], which is used in the traditional medicinal system of South Africa and Namibia to treat different forms of rheumatic disorders and back pain.

Rahmatullah, 2010a surved the six districts found that, the Kavirajes use 32 plants distributed into 23 families for treatment of rheumatoid arthritis. The Araceae and the Solanaceae families were the major contributors with three plants per family. Other families contributing two plants per family included Euphorbiaceae, Meliaceae, Piperaceae, Poaceae, and Rutaceae families. They also found that Leaves constituted the major plant part used (28.8%) followed by fruits (16.9%). Whole plant constituted 13.6% of total uses along with seeds, also at 13.6%. Barks constituted 8.5% of total uses.

The leaf of the Sheuli gach family name Verbenaceae and scientific name *Nyctanthes arbor tristis* L. is used for the Rheumatism, 15-20 leaves are boiled in 2 cups water and taken every morning and night for rheumatism. The whole plant of Gondho-vadule Scientific name *Paederia foetida* L. and family Rubiaceae is also used to treat Rheumatism. Leaf, stem and flower is boiled with oil and applied to affected area for rheumatism. Amrul shak including family Oxalidaceae is also used to treated for the Rheumatism In this case the plant is cooked as a vegetable and taken for 15 days for rheumatism. Go-roshun including family name is Liliaceae and Guloncho lota family name Menispermaceae is also applied for the Rheumatism (Rahmatullah et. al., 2010b)

Skin disease

Sida cordifolia L. the local name is Brela the family is Malvaceae plant is used for the skin disease in the kushtia region of Bangladesh. Usually the whole plant is used for the skin diseases. Powdered plant is mixed with camphor powder and applied to affected areas (Rahmatullah et. al., 2010b).

Curcuma longa L. family name, Zingiberaceae and local name Holud Rhizome is used too lighten up skin (improve skin texture), sprain. Raw rhizome is mixed with oil, applied to body followed by bathing to improve texture of skin. 4 parts rhizome, 1 part salt and 1 part lime are mixed thoroughly, warmed and applied to sprains for 2-3 days. Durba ghash the scientific name Cynodon dactylon including family Poaceae is used to treated Skin diseases,In this case whole plant is used 1 tola approximates 12.5g) of whole plant is mixed with 1 poa (local measure approximates 250g) sesame oil, heated and applied to affected areas as treatment for skin diseases (Rahmatullah et. al., 2010b).

Leprosy

Leprosy, otherwise known as Hansen's disease is a chronic disease caused by the bacteria *Mycobacterium leprae* and *Mycobacterium lepromatosis*. It is primarily a granulomatous disease of the peripheral nerves and mucosa of the upper respiratory tract. The disease is characterized by the formation of skin lesions. Kavirajes used *Terminalia arjuna* plants was used for the treatment of the Leprosy. Whole plants as well as plant parts like leaves, barks, roots and seeds were used for treatment.

In some cases, a single plant part (like bark of *Terminalia chebula* Retz.) was used. However, the Kavirajes also used combinations of two or more plant parts for treatment. A combination of leaves and roots of *Nerium indicum* Mill. was used by the Kavirajes of Netrakona district. A combination of leaves, roots, and seeds of *Cassia occidentalis* L. was used by the Kavirajes of Panchagarh district (Rahmatullah et. al, 2010;)

Visceral leishmaniasis

Isopropylquinolines, isolated from *Galipea longiflora inboligia* (Ritaceae), has also shown activity for visceral leishmanias (Pandey, 2009).

Cytotoxic effect

The leaves of *Euphorbia kamerunica* are toxic to rats (Ajibesin, 2002). The plant is also a known irritant having in vitro cytotoxic activities (Abo and Evans, 1981). Most of the impotant drug are of natural origin, as over 60% of approved drugs or those in late stages of development (during 1989-1995) are of natural origin (Crag et al., 1997). Some of the important antitumor agents derived from plants include paclitaxel, vincristine, and camptothecin. Furthermore, the broad reaching support and continuation of studies of plant extracts with implications in pancreatic cancer treatment are indicative of the continued role that natural products play in the drug discovery process (Schwarz et al.,2003. Lau et .al, 2009.

George et al., 2010, studied 56 extracts of 44 unique medicinal plants. The extracts were screened for cytotoxicity against the pancreatic adenocarcinoma cell line Panc-1, using a label-free biosensor assay. The top cytotoxic extracts identified in this screen were tested on two additional pancreatic cancer cell lines (Mia-Paca2 and Capan-1) and a fibroblast cell line (Hs68) using an MTT proliferation assay. Finally, one of the most promising extracts was studied using a caspase-3 colorimetric assay to identify induction of apoptosis. Crude extracts of *Petunia punctata*, *Alternanthera sessilis*, and *Amoora chittagonga* showed cytotoxicity to three cancer cell lines with IC50 values ranging between 20.3 - 31.4 µg/mL, 13.08 - 34.9 µg/mL, and 42.8 - 49.8 µg/mL, respectively. Furthermore, treatment of Panc-1 cells with *Petunia punctata* was shown to increase caspase-3 activity, indicating that the observed cytotoxicity was mediated via apoptosis. Only Amoora chittagonga showed low cytotoxicity to fibroblast cells with an IC50 value > 100 µg/mL. Based upon the initial screening work reported here, further studies aimed at the identification of active components of these three extracts and the elucidation of their mechanisms as cancer therapeutics are warranted.

Examples of clinically useful antitumor agents derived from plants include paclitaxel, vincristine, and camptothecin. Many of these plant-derived anticancer agents have been discovered through large-scale screening programs (Pezzuto 1997).

Medicinal Plants and its Molecular effect
Inhibition of Trancription factor and binding of DNA with Protein

For the gene expression transcription factor play an impotant role (Ji and Wong, 2006). Several TFs are indeed involved in inflammatory processes, including Nuclear Factor-kappa B (NF-kB) (Sebban and Courtois, 2006, Hayden et.al., 2004, Chen et.al.,2002, Monteleone et.al, 2004, Liao et.al., 2004, Loncar et.al.,2003), activator protein 1 (AP-1) (Wagner and Eferl, 2005, Yoshimura et.al., 2003, Hanada and Yoshimura, 2002,), signal transducer and activator of transcription (STATs) (Hodge et. al., 2005), cAMP response element binding protein (CREB) (Nakajima et.al., 2004) and GATA-1 factors (Masuda et.al., 2004, Hirasawa et.al., 2002).

All these TFs have been described to play a significant role in regulating expression of genes coding proinflammatory cytokines and in the pathogenesis of a large number of human disorders, particularly those characterized by a chronic inflammatory component (Hanada and Yoshimura, 2002). These include rheumatoid arthritis (Sato and Takayanagi, 2006, Nozaki et.al.,2006), chronic asthma (Adcock et. al., 2005, and Blease et.al., 2003), diabetes mellitus type 1 (Gray and De Meyts, 2005) and Crohn's disease (Stucchi et.al., 2006). With respect to anti-inflammatory activities, several plant-derived compounds exhibit significant effects (Tohda et.al., 2006, Wu et.al., 2004, Ahmed et.al., 2005, Salem, 2005, Fylaktakidou et.al., 2004,Nozaki et.al.,2006). Therefore, medicinal plants represent a potential source of molecules of significant relevance for developing novel drugs, especially designed for treating and/or controlling chronic inflammatory conditions.

Lampronti et al., 2008 screened Bangladeshi medicinal plants for their activity in inhibiting the interactions between nuclear factors and double stranded target oligonucleotides mimicking the NF-kB, AP-1, STAT-3, CREB and GATA-1 binding sites. They did a electrophoretic mobility shift assay (Borgatti et.al., 2003) . Extracts from several medicinal plants have been used, including *Emblica officinalis, Aegle marmelos, Moringa oleifera, Terminalia arjuna, Vernonia anthelmintica, Oroxylum indicum, Saraca asoka, Rumex maritimus, Lagerstroemia speciosa,* Red sandal, *Cuscuta reflexa, Argemone mexicana, Hemidesmus indicus, Polyalthia longifolia,*

Cassia sophera, *Paederia foetida*, *Hygrophilla auriculata*, *Ocimum sanctum* and *Aphanamixis polystachya*.

Antimirobial Activity

Leucas aspera found to contain sterols, fatty acids, lactones, long-chain compounds, aliphatic ketols, and phenols, which may have antifungal effects (Ahmed et al., 2010).

The component of the Garlic (*Allium sativum*) have a strong antimicrobial activity. It is utilized as a dietary component and as a substrate for the production of medicines (Baasinska and Kulasek, 2004). Limonoids present in the seed extract of Mahogany (*Swietenia mahagoni* Jacq) shown to have antimicrobial activity (Rahman et. al., 2009). For women the most common UTIs is leucorrhea, characterized by a whitish discharge from female genitalia. Microorganisms that causes leucorrhea included *Gardnerella vaginalis*, *Candida albicans*, *Chlamydia trachomatis* and *Trichomonas vaginalis* (Abauleth et al., 2006). In Bangladesh the prevalence rates of syphilis and gonorrhea among men was observed to be 4.1 and 7.7%, respectively (Alam et al., 2007).

Islam et al., 2010 studied the antimicrobial activity of chloroform soluble fraction *Allamanda cathartica* Leaves, methanolic soluble fraction *Curcuma zedoaria* of rhizome, carbon tetrachloride soluble fraction of *Callistemon citrinus* leaves and chloroform soluble fraction of *Stereospermu personatum* stem bark and showed significant antimicrobial activity ranging from 9-16 mm over all zone of inhibition in diameter. All the fractions were tested against different gram positive and gram negative bacteria and fungi to find out their antimicrobial activity using disc diffusion technique for bacteria and food poison method for fungi *Shigella dysenteriae*. *Bacillus cereus* (12 mm) Leaves' extract of *Allamanda cathartica* of chloroform soluble fraction, rhizome extract of *Curcuma zedoaria* of methanolic extract, leaves' extract of *Callistemon citrinus* of carbon tetrachloride soluble fraction and stem bark extract of *Stereospermu personatum* of chloroform soluble fraction showed significant antimicrobial activity ranging from 9-16 mm over all zone of inhibition in diameter.

The antibacterial activity of petroleum ether, carbon tetrachloride and chloroform soluble fractions of crude methanol extracts of nine indigenous plant species of Bangladesh was evaluated by the agar diffusion method. Kanamycin (30 µg/disc) was used as a standard antibacterial agent. The results indicated that all the nine plant species (not all partitionates) exhibited moderate to potent antibacterial activity against a wide variety of gram- positive and gram-negative bacteria at a concentration of 400µg/disc. Among them the carbon tetrachloride soluble fraction of whole plant extract of *Corriandrum sativum* (family-Apiaceae) revealed the highest antibacterial activity against Shigella boydii with zone of inhibition of 29 mm.

Rahman et al., 2008 studied the antimicrobial activity of methanol extracts of 17 plant species of Bangladesh was evaluated by the agar disc diffusion method. Among those, eight plant extracts exhibited potent antimicrobial activity against a wide variety of Gram positive and Gram negative bacteria and fungi at a concentration of 400 µg /disc.

Mahesh and Satish 2008, studied the methanol leaf extracts of *Acacia nilotica*, *Sida cordifolia*, *Tinospora cordifolia*, *Withania somnifer* and *Ziziphus mauritiana* showed significant antibacterial activity against *Bacillus subtilis*, *Escherichia coli*, *Pseudomonas fluorescens*, *Staphylococcus aureus* and *Xanthomonas axonopodis* pv. malvacearum and antifungal activity against *Aspergillus flavus*, *Dreschlera turcica* and *Fusarium verticillioides* when compare to root/ bark extracts. *A. nilotica* and *S. cordifolia* leaf extract showed highest antibacterial activity against *B. subtilis* and *Z. mauritiana* leaf extract showed significant activity against *X. axonopodis* pv. malvacearum. Root and leaf extract of *S. cordifolia* recorded significant activity against all the test bacteria. A. nilotica bark and leaf extract showed significant antifungal activity against *A. flavus, Ziziphus mauritiana* and *Tinospora cordifolia* recorded significant antifungal activity against *D. turcica* The methanol extract of *Sida cordifolia* exhibited significant antifungal activity against *F. verticillioides*.

The antibacterial activity of methanol extract from the root bark of Akanda (Calotropis gigantea L.) and its petroleum ether, chloroform and ethyl acetate fractions were investigated. Both of methanol extract and its chloroform fraction showed activity against Sarcina lutea, Bacillus megaterium and Pseudomonas aeruginosa (Alam et. al., 2008). Yasmin et. al., 2009 studied the antibacterial activity of petroleum ether, carbon tetrachloride and chloroform soluble fractions of crude methanol extracts of nine indigenous plant species of Bangladesh was evaluated by the agar diffusion method. Kanamycin (30 µg/disc) was used as a standard antibacterial agent. The results indicated that all the nine plant species (not all partitionates) exhibited moderate to potent antibacterial activity against a wide variety of gram- positive and gram-negative bacteria at a concentration of 400µg/disc. Among them the carbon tetrachloride soluble fraction of whole plant extract of Corriandrum sativum (family-Apiaceae) revealed the highest antibacterial activity against Shigella boydii with zone of inhibition of 29 mm.

4 CHAPTER FOUR: *IN VITRO* PROPAGATION AND CONSERVATION

The increasing demand for herbal medicines in recent years due to their fewer side effects in comparison to synthetic drugs and antibiotics has highlighted the need for conservation and propagation of medicinal plants. In vitro propagation has been successfully employed for the conservation of medicinal crop genetic resources, particularly with those crops which are vegetatively propagated (Ghani, 1998).

Tissue culture

Mass propagation of plant species through *in vitro* culture is one of the best and most successful examples of commercial application of plant tissue culture technology. Recently, there has been much progress in this technology for some medicinal plants.

Tissue culture propagation and its importance in conservation of genetic resources and clonal improvement have been described by many workers: Barz et al. (1977), Datta and Datta (1985), Kukreja et al. (1989) and Jusekutty et al. (1993).

Tissue culture has greatly enhanced the scope and potentiality of mass propagation by exploiting the regenerative behavior in a wide range of selected horticultural and agricultural plants including the medicinal ones (Roy et al. 1994, Thiruvengadam and Jayabalan 2000, Islam et al. 2001 and Jawahar et al. 2008).

P. foetida is a climbing twining shrub emitting a bad smell. It is an important gregarious medicinal plant found in Bangladesh Amin et.al, 2003 clonally propagated Nodal and shoot tip explants from field-grown mature plants of *P. foetida* were cultured on MS supplemented with BA and Kn at different concentrations (viz. 0.2, 0.5, 1.0, 2.0 and 5.0 mg/l) for proliferation of axillary shoots.

The leaves are rich in carotene and vitamin C, also contain a high amount of keto alcohol, keto compound and alkaloid. Leaf juice is astringent and used for treatment of diarrhea in children . Decoction of leaves dissolves vesical calculi and acts as diuretic. Leaves and roots are also regarded as tonic and stomachache and given to sick and convalescing patients; also used as remedies for diarrhoea, dysentery and rheumatic affections. Roots and barks are used as emetic and in the treatment of piles, inflammation of spleen and pain in chest and liver. Fruit is a specific against toothache (Ghani, 1998).

Boerhaavia diffusa L., commonly known as 'Punarnava' in

Bengali, is an important medicinal herb of the family Nyctaginaceae. The whole plant of B. diffusa is used as medicinal purpose because it contains different alkaloids and organic acids of medicinal importance. Biswas et al., 2009 clonally propagated *Boerhaavia diffusa* L explant on the MS medium. Multiple shoots were induced on MS fortified with 2.0 mg/l BAP and 0.2 mg/l NAA within 30 days of culture. Maximum (93%) explants produced multiple shoots with an average 12 shoots per culture after two successive subcultures at 14 days interval in the same medium.

Picrorhiza kurroa Royle ex. Benth. (Kour), a fast depleting high value medicinal plant belongs to Scrophulariaceae and is endemic to alpine Himalayan mountains (3500 - 5000 m asl).

Jan et al., 2009 developed *in vitro* regeneration method for the mass propagation of *Picrorhiza kurroa*, an endangered and highly valued medicinal plant without using cytokinin. Single medium combination favours best shoot as well as root formation. The highest seed germination rate was obtained when seeds were given cold treatment at 4ºC for ten days. Among the various strengths of MS and B_5 media (Gamborg et al. 1968), MS supplemented with 0.6 mg/l NAA showed the highest percentage of direct organogenesis.

Vitex negundo L., a perennial shrub belonging to the family Verbenaceae, is an important medicinal plant. It grows abundantly in St. Martin's Island and commonly known as Nishinda. Generally leaf is used for medicinal purpose but root, flower and fruit also have the medicinal values (Hasan 1982). Leaf of the plant contains essential oil, an alkaloid, nishindin. Stem and bark contain flavonoid glycosides. Leaves of nishinda very effectively reduce the inflammatoryswellings of joints in rheumatic attacks. Juice of fresh leaf removes fetid discharges and worms from ulcers. Flower oil is applied to sinuses andscrofulous sores. Root juice is tonic, expectorant and diuretic (Ghani, 1998)

Afroz et.al., 2008 established An efficient protocol was established for rapid and large scale propagation of woody aromatic medicinal plant *Vitex negundo* L. by *in vitro* shoot multiplication from shoot tips and nodal segments of mature plant. Of the four different growth regulators BA, Kn, GA3, NAA and coconut water, MS fortified with BA 1.0 mg/l was found to be the most effective for inducing multiple shoots from nodal explants.

Sen et. al., 2009 established an efficient *in vitro* plant regeneration protocol was developed for the medicinally potent plant species *Phyllanthus amarus* Schum. and Thonn. (Euphorbiaceae) using nodal segment as explant. Maximum multiplication of shoots (15.275±0.96) was achieved on Murashige and Skoog's medium supplemented with BAP (0.5 mg/l) after 3-4 weeks of inoculation.

Biswas et al., 2007, established An efficient protocol for *in vitro* propagation of *Abrus precatorius* L. through induction of indirect organogenesis in nodal segment derived callus tissue. Yellowish-green nodular callus was induced at the cut surface of the nodal segments cultured on MS fortified with 5.0 mg/l BAP and 0.5 mg/l NAA.

Abrus precatorius L. commonly known as 'Kunch' in Bengali is a deciduous woody climber of the family Fabaceae. It can be easily recognized by shiny scarlet coloured seeds with a black spot at one end. Since last long this plant species has been in use for its medicinal value (Kirtikar and Basu 1980, Biswas and Ghosh 1973). Different plant parts of this species contain various kinds of alkaloids such as glycerrhizin, precol, abrol, abrasine, abrin A and abrin B which impart its medicinal value (Joshi 2000, Ghani, 2003). The herbalists of Chittagong Hill Tracts (CHT) use seeds, leaves and roots of *A. precatorius* to induce abortion, pains and skin diseases. In CHT this medicinally important plant species is facing extinction due to indiscriminate collection, large scale deforestation and *Jhum* cultivation.

In nature the propagation of *A. precatorius* through seeds is difficult because of their hard seed coat - a trait which explains its sparse distribution. It is, therefore important to develop a protocol for *in vitro* propagation to save this medicinally important taxon from further depeletion of its population, at the same time to meet up the demand of the traditional medicine industry.

Ficus religiosa L. commonly known as 'Ashathwa' belongs to Moraceae is a large, widely branched tree with leathery, heart-shaped, long-tipped leaves on long slender petioles and purple fruits growing in pairs, grows both wild and planted throughout the Bangladesh (Ghani, 1998). The tree is found wild or cultivated nearly throughout India and is held sacred by Hindus and Buddhists. It is planted as an avenue or roadside tree. It grows fast and can be raised from seeds.

It can also be propagated by cuttings, but these do not establish so well as those of Ficus bengalensis. The fruits and tender leaf buds are occasionally eaten in times of scarcity. The fruits are eagerly devoured by birds. Extract of bark is antibacterial, astringent, relaxant an spasmolytic on smooth muscles and is used in diarrhoea, dysentery, gonorrhea, scabies and ulcers (Ghani, 1998). An aqueous extract of the bark shows anti-bacterial activity against Staphylococcus aureus and Escherichia coli. Leaves and young shoots are purgative and used in skin diseases. Seeds are cooling, alterative and laxative; taken for three days during menses, sterilizes women for long time. Ethanolic extract of bark is antiprotozoal, anthelmintic and antiviral (The Wealth of India 1956).

Hassan et.al., 2009 established A protocol for the mass propagation of the valuable medicinal plant *Ficus religiosa* L. (Moraceae) through in vitro culture using apical and axillary buds of young sprouts from selected plants. Best shoot induction was observed on MS basal medium supplemented with 0.5 mg/l BAP + 0.1 mg/l IAA, in which 78 per cent of the explants produced 16 shoots per culture.

Plants have always formed a rich source of modern drugs. The medicinal plants used by the Kavirajes need to be scientifically studied for phytochemical constituents and pharmacological activities towards discovery of lead compounds and more efficacious newer drugs.

Traditional medicinal practices used to be in vogue prior to the advent of modern allopathic medicine. While allopathic medicine to a certain extent diminished the importance of traditional medicinal practices, the former did not obliterate the latter. In fact traditional medicinal practices have a substantial number of adherents and have continued in practically all countries of the world as of this day. In recent years, traditional medicine is making a comeback because certain realizations have set in – the contribution of indigenous or traditional medicine in the discovery of new drugs (Balick and Cox, 1996.), the failure of modern drugs to cure all diseases, the emergence of modern drug-resistant organisms, the deleterious side-effects of a number of modern drugs, and last but not the least, a growing recognition among both patients and scientists that traditional medicine can also prove to be a successful way in the treatment of a number of diseases. Folk medicine forms one form of traditional medicinal practices in Bangladesh. Folk medicinal practitioners, known usually as Kavirajes, rely on simple preparations of medicinal plants to treat diseases. More often a single plant or plant part is used for treatment, although occasionally a complex combination of medicinal plants or plant parts are used. However, the preparation mode is usually simple and mainly consists of decoctions, macerations or pastes of whole plant or plant parts, which may be administered either orally or topically. In our ongoing ethnomedicinal surveys among the different tribes and in different regions of Bangladesh (Rahmatullah,*et al* 2010; Mollik,*et al* 2010; Hossan,*et al* 2010; Rahmatullah,*et al* 2010; Nawaz,*et al* 2009;Hanif, *et al* 2009; Hossan,*et al* 2009; Rahmatullah,*et al* 2009), it has been observed that any individual Kavirajes repertoire of medicinal plants vary considerably from another Kaviraj, who may be practicing in an adjoining area. This diversity of medicinal plants used by different Kavirajes makes it imperative to conduct ethnomedicinal surveys in as many areas of Bangladesh as possible to get a comprehensive view of the medicinal plants of Bangladesh. Kavirajes generally practice in the rural areas including small towns of the country. The objective of the present study was to conduct a survey on medicinal plant usage by Kavirajes of Barisal town in Barisal district of Bangladesh.

Khaton and Shaik

5. CHAPTER FIVE: PLANTS EXTRACTS AND METABOLITES

Ethanolic root extract of *Leucas aspera* root found to be antinociceptive, antioxidant and cytotoxic activity (Rahman et. al., 2007) and ethanolic leaf extracts shown to have potent and novel therapeutic agents for scavenging of Nitric oxide and the regulation of pathological conditions caused by excessive generation of NO and its oxidation product (Saha et. al., 2008).

An ethanolic *Acorus calamus* root extract are reported to exert sedative effects and potentiate hypnosis in vivo (Vohora et al., 1990; Zanoli et al., 1998). Root extract of *Acorus calamus* protected rats against acrylamide-induced neurotoxicity and reduced the incidence of paralysis (Shukla et al., 2002).

The essential oil (0.1% of the plant) extracted from Thankuni (*C. asiatica*) leaf contains monoterpenes, including bornyl acetate, a-pinene, which are reported to inhibit AChE (Miyazawa et al., 1997; Perry et al., 2000a; Ryan and Byrne, 1988). Alkaloids, hydrocotylin have been identified in *C. asiatica* (Duke and Ayensu, 1985),have the same acitivity. Triterpene and brahmoside shown to be present in the alcoholic extract of *C. asiatica* (Kapoor, 1990; Sakina and Dandiya, 1990). Extract of C. asiatica leaf to be sedative, antidepressant and potentially cholinomimetic in vivo (Sakina and Dandiya, 1990), appropriate to treat symptoms of depression and anxiety. Cognitive-enhancing effects have been reported from aqueous extract of *C. asiatica*, this effect was associated with an antioxidant mechanism in the CNS (Kumar and Gupta, 2002b).

Steroidal saponins are the main compounds found in a number of Asparagus species (Nwafor and Okwuasaba, 2002; Hayes et al.,2006a,b). Methanol extract of *Asparagus pubescens* inhibits albumin induced rat paw oedema (Nwafor and Okwuasaba, 2002). Roots of *Asparagus officinalis* contain flavonoids (Kartnig et al., 1985), oligosaccharides (Fukushi et al., 2000), amino acid derivatives (Kasai and Sakamura, 1981) and steroidal saponins (Shao et al., 1997).

A methanol extract of haritaki (*Terminalia chebula*) is reported to bind to NMDA and GABA receptors, but did not show anti-ChE activity (Dev, 1997); Another study showed an aqueous extract of *T. chebula* to be antioxidant (Naik et al., 2002).

The crude ethanolic stem bark extract of *Aphanamixis polystachya* (Meliaceae) has shown hepatoprotective activities (Gole et. al., 1993; Gole et. al., 1994; Gole et. al., 1995a; Gole et. al., 1995b; Gole et. al., 1997).

Mahogany *(Swietenia mahagoni* Jacq) seed extract so far has revealed that it contains triterpenoids possessing anti-platelet activity.(Ekimoto et. al., 1991). Methanolic extract of the dried ground seeds of mahogany *(Swietenia mahagoni* Jacq) has shown to have triterpenoids which been evaluated for anti-inflammatory, analgesic, and antipyretic activities (Ghos et. al., 2009)

Promising compounds from wo species from Rutaceae family are the 2-aryl and 2-alkylquinoline alkaloids isolated from the extracts of the stem bark, root bark and leaves of *Galipea longiflora* (Fournet et al. 1994) and lignans from *Zanthoxylum naranjillo* (Bastos et al. 1999), both are not reported in Bangladesh but similar compound can be found in some other plant of this family.

Some curcuminoids from *C. longa* rhizomes have shown associated with antioxidant and anti-inflammatory activities. Curcumin, a curcuminoid from *C. longa* rhizomesshown potential antioxidant activity (Das and Das, 2002; Miquel et al., 2002; Priyadarsini, 1997; Scartezzini and Speroni, 2000), anti-inflammatory (Miquel et al., 2002), neuroprotective (Rajakrishnan et al., 1999). Some compounds from *C. longa*, including curcumin, demethoxycurcumin, bisdemethoxycurcumin and calebin-A (and some of its synthetic analogues), shown antioxidant effect (Kim et al., 2001). An aqueous extract of *C. longa* demonstrated management of symptoms of cognitive-related disorders (antidepressant activity) in mice following oral administration (Yu et al., 2002). Components of ginger such as gingerol can inhibit the production of prostaglandins possibly more effectively that the arthritis drug domethicin.

Ashwogondha (*Withania somnifera* L.) is one of the most highly regarded herbs in Ayurvedic medicine. Steroid lactones found in *Ashwogondha* such as withanolides A-Y, glycowithanolides, dehydrowithanolide-R, withasomniferin-A, withasomidienone, withasomniferols A-C, withaferin A, withanone have been isolated from the root and leaf (Williamson, 2002). The phytosterols and sitoindosides VII–X alongwith the alkaloids ashwagandhine, ashwaghandhinine, cuscohygrine, anahygrine, tropine, pseudotropine, anaferine, isopelletierine, withasomine, visamine, somniferine, somniferinine, withanine, withaninine, pseudowithaninine and solasodine.

Different kind of secondary metabolites including triterpenoids, flavonol glycosides, anthocyanins and steroids has been isolated from *Clitoria ternatea Linn*. Its extracts possess a wide range of pharmacological activities including antimicrobial, antipyretic, anti-inflammatory, analgesic, diuretic, local anesthetic, antidiabetic, insecticidal, blood platelet aggregation-inhibiting and for use as a vascular smooth muscle relaxing properties. (Mukherjee, 2008)

This effects have a positive correlation on cholinergic activity in the CNS. A study investigating both the aerial parts and roots of *C. ternatea* showed alcoholic root extracts to be more effective in attenuating memory deficits in rats 2000). Enhanced memory retention following oral administration of C. ternatea root extract was associated with increased levels of ACh and choline acetyltransferase (ChAT) in rat brain, but any relationship with inhibition of AChE activity was not established, and cortical AChE activity was actually found to be increased (Taranalli and Cheeramkuzhy, 2000).

The extract of tulsi (*Ocimum sanctum*) leaves is hypoglycaemic, immuno-modulatory, anti stress, analgesic, antipyretic, anti-inflammatory, anti-ulcerogenic, antihypertensive, radio-protective, anti-tumour and antibacterial (Das and Vasdevan, 2006). The extract of tulsi leaves also used for the common cold for the baby.

The Chemicals compounds that have medicinal effects are (Lambert et. al., 1997) Alkaloids (compounds has addictive or pain killing or poisonous effect, Glycosides (use as heart stimulant or drastic purgative) (Srivastava et al., 1996) Tanins (used for gastro- intestinal problems like diarrhoea, dysentery, ulcer and for wounds and skin diseases), (Hopking et al., 2004) Volatile/essential oils (enhance appetite and facilitate digestion or use as antiseptic/insecticide and insect repellant properties), (Herbplace .com) Fixed oils (present in seeds and fruits could diminish gastric/acidity), (Natural health school) Gum- resins and mucilage (possess analgesic property that suppress inflammation and protect affected tissues against further injury and cause mild purgative), and (Ghani,1998) Vitamins and minerals (Fruits and vegetables are the sources of vitamins and minerals and these are used popularly in herbals (Ghani,1998).

They found a significant good result such as *P. longifolia* represents an extract displaying interesting selectivity in inhibiting only the NF-kB/DNA interaction, whereas for example *T. arjuna* demonstrated a very high activity in all TFs/DNA binding experiments. Both extracts derived from *E. officinalis* interfered preferentially with NF-kB/DNA interaction with lower selectivity than P. longifolia, while only one fraction of *A. marmelos* (pet. ether extract) was selective for GATA-1/DNA interaction. *Hemidesmus indicus* also appears to be an interesting selective extract, since it inhibits NF-kB/DNA interaction at 12 mg/ml, while it is active on the other TF/DNA interactions only when added at 50 mg/ml or higher concentrations. Both *A. polystachya* fractions proved a high, but not selective, activity particularly on the inhibition of NF-kB/ DNA interactions. *Moringa oleifera* and L. speciosa extracts were active on NF-kB, AP-1 and GATA-1 and inactive on other TFs. *Vernonia anthelmintica* demonstrated activity with NF-kB and GATA-1 only. *Saraca asoka* inhibited all TFs/DNA interactions even if at different concentrations. The aqueous fraction of R. maritimus was more active in inhibiting NF-kB, AP-1, GATA-1 and STAT-3 interactions than the methanolic extract, which interfered at intermediate concentrations with NF-kB/DNA interaction only. *Ocimum sanctum* was the only extract that selectively inhibited the STAT-3/DNA interaction. Finally, the remaining medicinal plant extracts were ineffective on all protein/DNA interactions.These data support the concept that medicinal plantextracts represent a potential source of compoundsexhibiting the ability to suppress the molecular interactionsbetween TFs and target DNA sequences. Further experiments [Gas Chromatography/Mass Spectrometry (GC/ MS), High-Performance Liquid Chromatography/Mass 309Spectrometry (LC/MS)] should be needed.

Mohammad S,2008, investigated, 35 locally used plants and tested to justify their existing bioactivities by the brine shrimp lethality bioassay. Out of the 35 plants used in the study 19 plants extracts were highly lethal to brine shrimp nauplii. This indicates that these plants contain potential bioactive compounds, which if properly and extensively studied, could provide many chemically interesting and biologically active drug candidates, including some with potential antitumor and antiproliferative properties.

Uddin et al., 2009 investigated, a total of 32 extracts representing 16 Bangladeshi plant (*Adiantum caudatum, Ammannia baccifera, Argemone mexicana, Blumea lacera, Clerodendron inerme, Ficus religiosa, Hygrophila auriculata, Limnophila indica* and *Mollugo pentaphylla*) including seven mangrove species (*Acrostichum aureum, Aegiceras corniculatum, Bruguiera gymnorrhiza, Cynometra ramiflora, Hibiscus tiliaceous, Pandanus foetidus* and *Xylocarpus moluccensis*) species from 16 plant families were screened for their cytotoxic activity against healthy mouse fibroblast and three human cancer cell lines (gastric, colon and breast cancer cells) Four extracts showed cytotoxic activity against all tested cell lines including the healthy cell line. The methanolic extract of *Adiantum caudatum* leaves displayed moderate cytotoxcicity (IC_{50} 1.23–1.88 $mgmL^{-1}$), whereas the aqueous extract from *Hibiscus tiliaceous* leaves showed. significantly lower IC_{50} values, especially against gastric (IC_{50} 0.25 $mgmL^{-1}$) and colon cancer cells (IC_{50} 0.8 $mgmL^{-1}$). However, the methanolic extract from *Blumea lacera* leaves showed the highest cytotoxicity (IC_{50} 0.01–0.08 $mgmL^{-1}$) against all tested cell lines among all

Medicinal plants in Bangladesh and its curative effects for disease

extracts tested in this study. Except *Cynometra ramiflora*, all of these plants have been used in traditional medicine of Bangladesh for the treatment of various diseases such as cancer, inflammation or infectious diseases

Table 18. LC50 of methanol extracts of some medicinal plants of Bangladesh.

Plant (Family)	Uses in traditional medicine	RMC	RPPS	LC50 Reference (µg/ml)
Aglaia roxburghiana (Meliaceae)	Dysentry, leucoderma, leprosy, fever, thirst, tumors, vomiting.6	24, 25-epoxy-29-norcycloartan-3-ol, 29-norcyclorart-23-ene-3, 25-diol, 24,25-epoxy-29-nor-24-cycloarten-3β-ol, roxburghiline, hydroxyroxburghiline, aglaroxin-A, roxhurghiadiol A.10	Terpenoids, alkaloids	11.66 ± 1.10
Amoora rohituka (Meliaceae)	Cancer, tumours, spleen and liver disease, rheumatism.7	6b,7b-epoxyguai-4-en-3-one, 6b,7b-epoxy-4b,5-dihydroxygualane,11 stigmasta-5,24(28)-dien-	Terpenoids	5.95 ± 0.94 Mohammad S,2008

Plant (Family)	Uses in traditional medicine	RMC	RPPS	LC50 Reference (µg/ml)
		3β-O-β-D-glucopyranosyl-O-α-Lrhamnopyranoside, 12 7-keto-octadec-cis-l 1-enoic acid.13		
Ficus indica (Moraceae)	Relieve toothache, rheumatism, lumbago, inflammations, diarrhoea, dysentry, vomitting, biliousness.7	Bengalenoside, leucoanthocyanidins, leucoanthocyanin glycoside, betasitusterol glycoside, mesoinositol, friedelin, beta-sitosterol, qurecetin-3-galactoside and rutin, tiglic acid ester of gamma-tarxerol, cyanidin rhamnoglycoside, ficusin and bergaptin.	Steroids, terpenoids, phenolics	17.67 ± 1.17 Mohammad S,2008
Solamum indicum	Astringent,	Isoanguivine, protodioscin,	Steroid	4.42 ±

Medicinal plants in Bangladesh and its curative effects for disease

Plant (Family)	Uses in traditional medicine	RMC	RPPS	LC50 Reference (µg/ml)
(Solanaceae)	carminative, cardiac tonic, aphrodisiac	solasonine and solamargine, indioside A-E.	s	0.67 Mohammad S,2008
Terminalia bellirica (Combretaceae)	Hepatitis, breathing problem, coughs, eye disease, constipation purgative.7	Cardenolide,43 2-dotriacontanol, bellericagenin B, bellericaside B, termilignan, bellericagenin A, bellericaside A, thannilignan, 9-tritriacontanone, 10 sitosterol, gallic aad, ellagic, acid, ethyl gallate, galloyl glucose, mannitol, glucose, galactose, fructose and rhamnose, bellerlc acid, bellericoside, arjungenin, arjunglycoside.44	Terpenoids, phenolics, steroids,	3.62 ± 1.35 Mohammad S,2008
Terminalia chebula	Indigestion, constipation	Terflavin A, chebulagic acid, chebulinic acid,	Phenolics	8.85 ± 0.38

Plant (Family)	Uses in traditional medicine	RMC	RPPS	LC50 Reference (µg/ml)
(Combretaceae)	jaundice, piles, painful menstruation, asthma, colic, as diuretic and cardiotonic.6	corilagin, 2α-hydroxymicromeric acid, luteolic acid, 12-oleanene-2,3,19,23,28-pentol, terchebin, terchebulin, terflavin D, terfalvin B, 1,3,6-trigalloyl glucose.		Mohammad S,2008
Xylocarpus granatum (Meliaceae)	Dysentery, diarrhea, and other abdominal problems.7	xyloccensins O-P, xyloccensins Q-V, 45 Xyloccensin L, 46 xyloccensin K, 47 xyloccensin IJ. 48	Terpenoids	6.81 ± 0.22

6. CHAPTER SIX: CONSERVATION

In terms of conservation, 13 endemic species can be considered endangered since the only organ used in the majority of them is the root (*Eryngium agavifolium, Hypochaeris pampasica, Senecio uspallatensis, Trichocline plicata, Trichocline sinuata, Berberis lilloana, Berberis grevilleana, Adesmia inflexa, Polygala stenophylla,* and *Valeriana ferax*). Some species are very rare with a restricted distribution (*Cyperus spectabilis* var. *jujuyensis, Chiliophyllum densifolium, Mutisia saltensis, Senecio pogonias, S. uspallatensis, Trichocline plicata, Senna kurtzii, Gentianella imberbis, G. parviflora, Sphaeralcea philippiana, Siphoneugena occidentalis*) while others need priorities in conservation practices due to the progressive destruction of their habitats (e.g. *Polylepis australis, Condalia microphylla, Lippia* spp.).

Many types of action can be taken in favour of the conservation and sustainable use of medicinal plants. Some of these are undertaken directly at the places where the plants are found, while others are less direct, such as some of those relating to commercial systems, *ex situ* conservation and bioprospecting. In the latter cases, actions taken will not lead to *in situ* conservation unless they feed back to improvements in the field. Probably the single most important role for medicinal plants in biological conservation is their 'use' to achieve conservation of natural habitats more generally. This stems from the special meanings that medicinal plants have to people, related to the major contributions that they make to many people's lives in terms of health support, financial income, cultural identity and livelihood security (Hamilton, 2004).

Ethnobotanical data helps to identify medicinal plant species, which in turn can lead to sustainable cultivation and preservation of endangered medicinal plants (Rahmatullash et al., 2010). In the South Asian countries Bangladesh have a large number of valuable medicinal plants naturally growing mostly in fragile ecosystems that are predominantly inhabited by rural poor and indigenous community (Kharkil et.al., 2003).

Conservation of medicinal plants is very important, partly because many have not yet been studied. Also most of the rural and urban people don't know which plant is good or not so usually people don't take care the plant or some times destroy them. So for conservation the first step is to identify the medicinal plant properly so that general people can realized about the useful plants. If these plants become too rare or expensive, however, people in the developing world may lose access to their primary medicines. The Garden supports conservation, economic development, and education programs in tropical countries. One goal of such programs is to help people use plants and other natural resources sustainably, so they continue to be available in the future. Medicinal plant species which are rare or endangered or threatened should be identified and their ex-situ conservation, may be attempted in the established gardens, plantations and other areas.We also need to develop gene bank by the help of Government so that we should properly store the germless of all medicinal plants.Micropropagation is a good technique for the rapid multiplication of the plant. By using the micropropagation technique it is possible to get a disease free and elite medicinal plant.

Biotechnology for the conservation of the Medicinal plants

Biotechnological technique play an important for the selection ,rapid multiplication and conservation of the rare genotypes of medicinal plants. Tissue culture and micropropagation actively helped the production of high-quality plant-based medicine.. In-vitro production of secondary metabolites in plant cell suspension cultures has been reported from various medicinal plants. Bioreactors are the key step towards commercial production of secondary metabolites by plant biotechnology.

7. CHAPTER SEVEN: REGISTRATION & PROPERTY RIGHTS

WHO has published guidelines for the assessment of herbal medicines taking into account long and extensive usage of them (WHO, 1999). These guidelines should encourage developing countries to relax some of the current regulations to be realistic in recognising the role of traditional medicines in the health care delivery of their countries.

Attempt should be made to identify traditional health practices and knowledge relating to process and products of medicinal plants and the information should be digitalized and put on computer. Wherever possible patents may be obtained for the process and products of medicinal plants. The vital question of property right in developing countries for the use of know-how and genetic resources in the development of modern drugs has to be discussed and a final solution will be found.

8. CHAPTER EIGHT: RESEARCH AND DEVELOPMENT

For most of human history, plants and a few fungi were the only medicines. People discovered medicinal plants by trial and error; that is, by tasting them to see what happened! Plant-based medicines are often less potent than modern drugs, but studies prove many to have real value. Some do not, however, and a few are so toxic that they may do more harm than good. Because research facilities are limited, most locally used plants have never been studied at all. Over four billion people in developing countries rely on traditional medicines. They often cannot be sure whether a medicine is safe or effective.

For getting the economic value from medicinal plant it is necessary intensive research. The research and development is necessary in each step from plant material collection to utilization. It is also necessary to categorize medicinal species into : (a) those which are of proven medicinal value as per scientific parameters, (b) Characterization of all medicinal plant that are available in each district (c) It is also very important to molecular characterization of the valuable compound that are present in the medicinal plant.(d) It is also necessary to know the molecular mechanism how the metabolite work for the specific disease. (e) biochemical, physiological and genetic studies on the herbal plants themselves, in order to distinguish between those originating from different habitats, or to improve the known medicinal quality of the indigenous plant.

Research investigations need to concentrate on the first two categories on the following

aspects:-

• Evolving and optimising the most appropriate technologies for conservation, especially for endangered or endemic species and molecular methods for characterisation. If possible to develop molecular marker aassisted breeding and characterization.

• Detailed studies on life cycle and breeding behaviour, taxonomy, seed biology

• Population and habitat viability studies

• Optimising appropriate methods for post-harvest handling, processing and storage.

• Investigation on quality control standardization and shelf life of raw materials and finished products.

For Bangladeshi medicinal plants it is also needed to develop a web based information so scientist or local can people can easily access the information.

9. CHAPTER NINE: CONCLUSION

In the world, 40,000-70,000 medicinal plants species are estimated and still there is a lot of traditional knowledge that has not yet been really explored (Verpoorte, 2007). In Bangladesh, the composition of the native ethnopharmacopoeia has increased considerably specially with the contribution of detailed ethnobotany studies carried out in different aborigenous and rural communities where their healthcare systems continue to rely on their traditional plant-based medicines. More than 200 medicinal species reported in this review, which represent near the 40 % of the total Bangladeshi flora, demonstrate that we have a good source of promising plants that should be studied in detail. This value will be still greater if we also consider the introduced medicinal species.

10. CHAPTER TEN: REFERENCES

Abauleth, R., S. Boni, A. Kouassi-Mbengue, J. Konan & S. Deza. 2006. Causation and treatment of infectious Leucorrhea at the Cocody University Hospital (Abidjan, Côte d'Ivoire). Sante 16(3):191-195.

Abo, K., and Evans, F.J. 1981. The composition of a mixture of Ingol Esters from Euphorbia kamerunica, Planta Medica 43(12): 392-395.

Acuna UM, Atha DE, Ma J, Nee MH, Kennely EJ. Antioxidant capacities of ten edible North American plants. Phytother Res 2002;16(1):63 – 5.

Adcock IM, Ito K, Barnes PJ. Histone deacetylation: an important mechanism in inflammatory lung diseases. COPD 2005;2:445-55.

Adcock IM, Ito K, Barnes PJ. Histone deacetylation: an important mechanism in inflammatory lung diseases. COPD 2005;2:445-55.

Afroz F, Hassan AKMS, Bari LS, Sultana R, Munshi JL, Jahan M A A and Khatun R. In Vitro Regeneration of Vitex negundo L., a Woody Valuable Medicinal Plant through High Frequency Axillary Shoot Proliferation. *Bangladesh J. Sci. Ind. Res.* 43(3), 345-352, 2008

Agbovie, T., Amponsah, K., Crentsil, O.R., Dennis, F., Odamtten, G.T. and Ofusohene-Djan, W. 2002. Conservation and sustainable use of medicinal plants in Ghana, Ethnobotanical Survey, UNEPWCMC, Cambridge, UK.

Ahmed F, Sadhu S K, Ishibashi M 2010. Search for bioactive natural products from medicinal plants of Bangladesh. *J Nat Med* 64:393–401.

Ahmed F., Sadhu S. K. and Ishibashi M. (2010) Search for bioactive natural products from medicinal plants of Bangladesh,

J Nat Med 64:393-401.

Ahmed S, Anuntiyo J, Malemud CJ, Haqqi TM. Biological basis for the use of botanicals in osteoarthritis and rheumatoid arthritis: a review. Evid Based Complement Alternat Med 2005;2:301-8.

Ahmed S, Anuntiyo J, Malemud CJ, Haqqi TM. Biological basis for the use of botanicals in osteoarthritis and rheumatoid arthritis: a review. Evid Based Complement Alternat Med 2005;2:301-8.

Ajibesin, K.K., Bala, D.N., Ekpo, B.A.J. and Adesanya, S.A. 2002. Toxicity of some plants implicated poisons in Nigerian ethnomedicine to rats, Nig. J Nat Prod Med 6: 7-9.

Akihisa, T., Kimura, Y., Kokke,W.C.M.C., Itoh, T., Tamura, T., 1996. Eight novel sterols from the roots of Bryonia dioica Jacq. Chemical & Pharmaceutical Bulletin 44, 1202-1207.

Akter R, Hasan SMR, Siddiqua SA, Majumder MM, Hossain MM, Alam MA, Haque S, Ghani A 2008: Evaluation of Analgesic and Antioxidant Potential of the Leaves of *Curcuma alismatifolia* Gagnep, *Stamford Journal of Pharmaceutical Sciences* 1(1):3-9.

Akter R, Hasan SMR, Hossain MM, Jamila M, Mazumder MEH and Rahman S 2009: In Vitro Antioxidant and In Vivo Antidiarrhoeal Activity of Hydromethanolic Extract of *Xanthium Indicum* Koenig. Leaves, *European Journal of Scientific Research* 33(2): 305-312.

Alam MA, Habib M R, Nikkon F, Rahman M and Karim M R.Antimicrobial Activity of Akanda (Calotropis gigantea L.) on Some Pathogenic Bacteria. *Bangladesh J. Sci. Ind. Res.* 43(3), 397-404, 2008

Alam, N., M. Rahman, K. Gausia, M.D. Yunus, N. Islam, P. Chaudhury, S. Monira, E. Funkhouser, S.H. Vermund & J. Killewo. 2007. Sexually transmitted infections and risk factors among truck stand workers in Dhaka, Bangladesh.Sexually Transmitted Diseases 34(2):99-103.

Amin M. N., Rahman M. M. and Manik M. S. In vitro Clonal

Propagation of Paederia foetida L.- A Medicinal Plant of Bangladesh Plant Tissue Cult. 13(2) : 117-123, 2003 (December) PTC

Animesh B, 2007, In vitro Propagation of *Abrus precatorius L. -* A Rare Medicinal Plant of Chittagong Hill Tracts *Plant Tissue Cult. & Biotech. 17(1): 59-64,* PTC&B

Anonymous. Medicinal plants in world market. Available at http://www.mapbd.com/Start.htm; 2008a.

Atta-ur-Rahman (1988) Studies in Natural Products Chemistry, Elsevier, Amsterdam, The Netherlands.

Baasinska B, Kulasek G 2004: Garlic and its impact on animal and human health. *Medycyna Weterynaryjna* 60 1151-1155.

Bala,V.,Manyam, M.D., 1999. Dementia in ayurveda. The Journal of Alternative and Complementary Medicine 5, 81–88.

Balandrin, M.F., J.A. Klocke, E.S. Wurtele and W.H. Bollinger, 1985. Natural plant chemicals: *corniculata*:Sources of Industrial and Medicinal materials. plant. My. Science., 1: 72-78.

Balick, J.M. and P.A. Cox, 1996. Plants, People and Culture: the Science of Ethnobotany, Scientific American Library, New York, pp: 228.

Bandaranayake WM (2002) Bioactivities, bioactive compounds and chemical constituents of mangrove plants. Wetl Ecol Manag 10:421–452

Banglapedia 2006, the National Encyclopedia of Bangladesh edited by Professor Sirajul Islam, published by Asiatic Society of Bangladesh.

Bastos JK, Albuquerque S, Silva MLA 1999. Evaluation of the trypanocidal activity of lignans isolated from the leaves of *Zanthoxylum naranjillo. Planta Med 65*: 541-544.

Betti, J.L. 2004. An ethnobotanical study of medicinal plants among the Baka Pygmies in the Dja Biosphere Reserve, Cameroon, African Study Monographs 25(1): 1-27.

Bharghava SK (1984) Effects of plumbagin on reproductive function of male dog, Ind. J. Exp. Biol. 22 : 153-156.

Bhowmik S, Chowdhury SD , Kabir M. H and Ali MA Chemical composition of some medicinal plant products of indigenous origin. *The Bangladesh Veterinarian* (2008) 25(1) : 32 – 39

Bhowmik S, Chowdhury SD, Kabir MH, and Ali MA 2008: Chemical composition of some medicinal plant products of indigenous origin. *The Bangladesh Veterinarian* 25(1): 32 – 39.

Bhowmik S, Chowdhury SD, Kabir MH, and Ali MA 2008: Chemical composition of some medicinal plant products of indigenous origin. *The Bangladesh Veterinarian* 25(1): 32 – 39.

Bio-ecological Zones of Bangladesh. IUCN-The World Conservation Union, Bangladesh Country Office. 2003.

Biswas A, M A. Bari M A, Roy M and Bhadra SK, 2009 Clonal Propagation Through Nodal Explant Culture of Boerhaavia diffusa L. - A Rare Medicinal Plant. *Plant Tissue Cult. & Biotech.* 19(1): 53-59, 2009 (June)

Biswas A, Roy M, MAB Miah and Bhadra SK. In vitro Propagation of Abrus precatorius L. - A Rare Medicinal Plant of Chittagong Hill Tracts Plant .Tissue Cult. & Biotech. 17(1): 59-64, 2007 (June)

Blease K, Lewis A, Raymon HK. Emerging treatments for asthma. Expert Opin Emerg Drug 2003;8:71–81.

Blease K, Lewis A, Raymon HK. Emerging treatments for asthma. Expert Opin Emerg Drug 2003;8:71–81.

Borgatti M, Lampronti I, Romanelli A, Pedone C, Saviano M, Bianchi N, et al. Transcription factor decoy molecules based on a peptide nucleic acid (PNA)-DNA chimera mimicking Sp1 binding sites. J Biol Chem 2003;278:7500–9.

Borgatti M, Lampronti I, Romanelli A, Pedone C, Saviano M, Bianchi N, et al. Transcription factor decoy molecules based on a peptide nucleic acid (PNA)-DNA chimera mimicking Sp1 binding sites. J Biol Chem 2003;278:7500-9.

Brien, S., G.T. Lewith and G. McGregor, 2006. Devil's Claw (Harpagophytum procumbens) as a treatment for osteoarthritis: a review of efficacy and safety. Journal of Alternative and Complementary Medicine, 12: 981-993.

Chatterjee A, Das B, Adityachaudhary N, Dabkirtaniya S (1980). Note on the insecticidal of the seeds of *Jatropha gossypifolia* Linn. Indian J. Agric. Sci. 50: 637-638.

Chen F, Demers LM, Shi X. Upstream signal transduction of NF-kappaB activation. Curr Drug Targets Inflamm Allergy 2002;1:137-49.

Chen F, Demers LM, Shi X. Upstream signal transduction of NF-kappaB activation. Curr Drug Targets Inflamm Allergy 2002;1:137-49.

Chinese Pharmacopoeia Commission, 2005. The Pharmacopoeia of the People's Republic of China, vol. 1. Chemical Industry Press, Beijing.

Choi, E.-M., Hwang, J.-K., 2003. Investigation of anti-inflammatory and antinociceptive activities of Piper cubeba, Physalis angulata and Rosa hybrida. Journal of Ethnopharmacology 89, 171-175.

Chopin, J., Dellamonica, G., Bouillant, M.L., Besset, A., Popovici, G., Weissenböck, G., 1977. C-glycosylflavones from Avena sativa. Phytochemistry 16, 2041-2043.

Chopra RN, Nayar SL, Chopra IC 1956: *Glossary of Indian Medicinal Plants* CSIR, New Delhi, India pp. 1-259.

Chowdhury AR, 2010: A Survey of Medicinal Plants Used by Kavirajes of Barisal. *American-Eurasian Journal of Sustainable Agriculture* 4(2): 237-246.

Chowdhury, M.S.H., Koike, M., Muhammed, N., Halim, M.A., Saha, N. and Kobayashi H. 2009. Use of plants in healthcare: A traditional ethno-medicinal practice in southeastern rural areas of Bangladesh. *International Journal of Biodiversity Science and Management, United Kingdom* 5(1): 41-51.

Cragg GM, Newman DJ, Snader KM: Natural Products in Drug Discovery and Development. Journal of Natural Products 1997, 60(1):52-60.

Croat, T.B. 1979. The distribution of Araceae. In: Larsen, K. Holm-Nielsen, L.B. (eds), *Tropical Botany*. Academic Press, London. Pp. 291-308.

Croat, T.B. 1994. Taxonomic status of neotropical Araceae. *Aroideana* 17: 33-60.

D. N. Tewari, 2000, Report of the Task Force on Conservation & Sustainable use of Medicinal Plants

Das KC, Das CK. Curcumin (diferuloylmethane), a singlet oxygen (O-1(2)) quencher. Biochem Biophys Res Commun 2002;295(1):62 –6

Das SK, Vasudevan DM 2006: Tulsi: the Indian holy power plant. *Natural Product Radiance* 5 279-283

Denner, S.S., 2007. A review of the efficacy and safety of devil's claw for pain associated withmdegenerative musculoskeletal diseases, rheumatoid, and osteoarthritis. Holistic NursingmPractice, 21: 203-207.

Dev S. Ethnotherapeutics and modern drug development: the potential of Ayurveda. Curr Sci 1997;73(11):909 –28.

Didry N, Dubrevil L and Pinkas M (1994) Activity of anthraquinonic and naphthoquinonic compounds on oral bacteria, Die Pharmazie 49 : 681-683.

Ekimoto H, Irie Y, Araki Y, Han GQ, Kadota S, Kikuchi T. (1991) Platelet aggregation inhibitors from the seeds of Swietenia mahagoni: inhibition of in vitro and in vivo platelet- activating factor-induced effects of tetranortriterpenoids related to swietenine and swietenolide. Planta Med; 57:56-8.

Fingl E (1980). Laxatives and cathartics. In The Pharmacological Basis of Therapeutics, 6th edn, Gilman AG, Goodman LS, Gilman A (eds). Macmillan: New York, pp. 1002-1012

Fournet A, Barrios AA, Munoz, V, Hocquemiller R, Roblot F,Cavé A, Richomme P, Bruneton J 1994. Antiprotozoal activity of quinoline alkaloids isolated from *Galipea longiflora*, a Bolivian plant used as a treatment for cutaneous leishmaniasis. *Phytother Res 8*: 174-178.

Fukushi, E., Onodera, S., Yamamori, A., Shiomi, N., Kawabata, J., 2000. NMR analysis of tri- and tetrasaccarides from Asparagus.Magnetic Resonance in Chemistry 38, 1005–1011.

Fylaktakidou KC, Hadjipavlou-Litina DJ, Litinas KE, Nicolaides DN. Natural and synthetic coumarin derivatives with anti-inflammatory/antioxidant activities. Curr Pharm Des 2004;10:3813-33.

Fylaktakidou KC, Hadjipavlou-Litina DJ, Litinas KE, Nicolaides DN. Natural and synthetic coumarin derivatives with anti-inflammatory/antioxidant activities. Curr Pharm Des 2004;10:3813-33.

Gao T., Yaoa H., Songa J., Liu C., Zhua Y., Ma X., Pang X. and Chen S. (2010) Identification of medicinal plants in the family Fabaceae using a potential DNA barcode ITS2 Journal of Ethnopharmacology *130 (2010) 116–121*

George S, S V Bhalerao S V , E A Lidstone E A , Ahmad I S, A Abbasi A , Cunningham B T, Watkin K L 2010. Cytotoxicity screening of Bangladeshi medicinal plant extracts on pancreatic cancer cells. *BMC Complementary and Alternative Medicine* , 10:52

Ghani A 2000: *Medicinal Plants of Bangladesh*: Chemical

Constituents and Uses, Asiatic Society of Bangladesh, Dhaka.

Ghani A.(1998), Medicinal plants of Bangladesh: Chemical constituents and uses. Asiatic society of Bangladesh, Dhaka.

Ghosal, S., Lal, J., Srivastava, R., Bhattacharya, S.K., Upadhyay, S.N., Jaiswal, A.K., Chattopadhyay, U., 1989. Bioactive phytosterol conjugates. Part 7. Immunomodulatory and CNS effects of sitoindosides IX and X, two new glycowithanolides from Withania somnifera. Phytotherapy Research 3, 201-206.

Ghosh S, Besra SE, Roy K, Gupta JK, Vedasiromoni JR. Pharmacological effects of methanolic extract of Swietenia mahagoni Jacq (meliaceae) seeds. Int J Green Pharm 2009;3:206-10

Gilani, A.H. and A.U. Rahman, 2005. Trends in ethnopharmacology. Journal of Ethnopharmacology, 100: 43-49.

Gill, L.S. 1988. Taxonomy of flowering plants, Africana-Fep Publishers Ltd., Nigeria.

Gole MK, Biswas T, Ghoshal J, Dasgupta S. Phytotherapy: A potent therapeutic source of liver disorder remedy. Proceedings of the 35th World Congress on Natural Medicines, 14-16 March, SV University, Tirupati, India, 1997: 98.

Gole MK, Biswas T, Ghoshal J, Dasgupta S. Role of antihepatotoxic natural product extract as biological antioxidant. Indian J Physiol Allied Sci 1994; 48: 2-3.

Gole MK, Chatterjee A, Ghoshal J, Dasgupta S. Antihepatotoxic constituents from medical plants: A systematic study. Proceedings of the First Congress of the Federation of Indian Physiological Societies, 1-3 March, 1995, DIPAS, New Delhi, India, 1995b: 123.

Gole MK, Ghoshal J, Dasgupta S. Antioxidant activity of *Aphanamixis polystachya* on carbon tetrachloride induced hepatic injury. Int J Toxicol Occ Env Health (Abstracts) 1993: 10.

Gole MK, Mishra M, Dasgupta S, Banerjee S, Ghoshal J. Radionuclide imaging in the evaluation of hepatoprotective activity of a natural product. Ind J Nucl Med 1995a; 10: 48-49.

Goyal, R.K., Singh, J., Harbans, L., 2003. Asparagus racemosus— an update. Indian Journal of Medical Sciences 57, 408-414.

Gray SG, De Meyts P. Role of histone and transcription factor acetylation in diabetes pathogenesis. Diabetes Metab Res Rev 2005;21:416-33.

Gray SG, De Meyts P. Role of histone and transcription factor acetylation in diabetes pathogenesis. Diabetes Metab Res Rev 2005;21:416-33.

Hamilton A. C. 2004. Medicinal plants, conservation and livelihoods. *Biodiversity & Conservation* 13: 1477- 1517.

Hanada T, Yoshimura A. Regulation of cytokine signaling and inflammation. Cytokine Growth Factor Rev 2002;13:413-21.

Hanada T, Yoshimura A. Regulation of cytokine signaling and inflammation. Cytokine Growth Factor Rev 2002;13:413-21.

Hartwell JL (1996). Plants used against cancer, a survey. Lloydia. 32: 153-205.

Hasan SMR, Hossain MM, Akter R, Jamila M, Mazumder MEH and Rahman S 2009a: DPPH Free Radical Scavenging Activity of some Bangladeshi Medicinal Plants, *Journal of Medicinal Plant Research*, 3(11):875-879.

Hasan SMR, Hossain MM, Akter R, Jamila M, Mazumder MEH, Alam MA, Faruque A, Rana S, Rahman S 2010: Analgesic Activity of the Different Fractions of the Aerial Parts of *Commelina benghalensis* Linn. *International Journal of Pharmacology* 6(1): 63-67.

Hasan SMR, Jamila M, Majumder MM, Akter R, Hossain MM, Mazumder MEH, Alam MA, Jahangir R, Rana MS, Arif M and Rahman S 2009b: Analgesic and Antioxidant Activity of the Hydromethanolic Extract of *Mikania Scandens* (L.) Willd. Leaves,

American Journal of Pharmacology and Toxicology 4(1): 1-7.

Hassan A KM S, Afroz F, Jahan MA A and Khatun R.In vitro Regeneration through Apical and Axillary Shoot Proliferation of Ficus religiosa L. A Multi-purpose Woody Medicinal Plant .Plant Tissue Cult. & Biotech. 19(1): 71-78, 2009 (June)

Hayden MS, Ghosh S. Signaling to NF-kB. Genes Dev 2004; 18:2195-224.

Hayden MS, Ghosh S. Signaling to NF-kB. Genes Dev 2004;18:2195-224.

Hiller, K., Melzig, F.M., 2006. Lexikon der Arzneipflanzen und Drogen. Spektrum

Hirasawa R, Shimizu R, Takahashi S, Osawa M, Takayanagi S, Kato Y, et al. Essential and instructive roles of GATA factors in eosinophil development. J Exp Med 2002;195:1379-86.

Hirasawa R, Shimizu R, Takahashi S, Osawa M, Takayanagi S, Kato Y, et al. Essential and instructive roles of GATA factors in eosinophil development. J Exp Med 2002;195:1379-86.

Hodge DR, Hurt EM, Farrar WL. The role of IL-6 and STAT3 in inflammation and cancer. Eur J Cancer 2005;41:2502-12.

Hodge DR, Hurt EM, Farrar WL. The role of IL-6 and STAT3 in inflammation and cancer. Eur J Cancer 2005;41:2502-12.

Hopking A.(2004). The glimpse over the history of Herbalism. Godshaer Herbalist advanced botanical center of medicine.www.godshaer.co.uk/history.htm

Hossain M.Z. et al. (2003). People's awareness about medicinal values of plants and prospect in Bangladesh. BRAC, Research and Evaluation division, Dhaka, Bangladesh.

Hossan MS, Hanif A, Agarwala B, Sarwar MS, Karim M, Taufiq-Ur-Rahman M, Jahan R, Rahmatullah M 2010: Traditional Use of Medicinal Plants in Bangladesh to Treat Urinary Tract Infections and Sexually Transmitted Diseases, *Ethnobotany Research &*

Applications 8:061-074.

http://www.life.umd.edu/

Ishibashi M, Arai MA (2009) Search for bioactive natural products targeting cancer-related signaling pathways. J Synth Org Chem 67:1094-1103ience, 228: 1154-1160.

Ishibashi M, Ohtsuki T (2008) Studies on search for bioactive natural products targeting TRAIL signaling leading to tumor cell apoptosis. Med Res Rev 28:688-714

Islam M R, Ahamed R, Rahman M O, Akbar MA, Amin MA, Alam K D and Lyzu F, 2010. In Vitro Antimicrobial Activities of Four Medicinally Important Plants in Bangladesh *European Journal of Scientific Research* Vol.39 No.2 (2010), pp.199-206

Jan A, Phalisteen S, Thomas GT and Shawl A S . Auxin Prompted Improved Micropropagation Protocol of Picrorhiza kurroa: An Endangered Medicinal Plant. *Plant Tissue Cult. & Biotech*. 19(2): 161-167, 2009 (December)

Ji H, Wong WH. Computational biology: toward deciphering gene regulatory information in mammalian genomes. Biometrics 2006;62:645-63.

Ji H, Wong WH. Computational biology: toward deciphering gene regulatory information in mammalian genomes. Biometrics 2006;62:645-63.

Kapoor LD. Handbook of Ayurvedic medicinal plants. Boca Raton (FL): CRC Press; 1990.

Kartnig, T., Gruber, A., Stachel, J., 1985. Flavonoid pattern from Asparagus officinalis. Planta Medica 51, 288.

Kasai, T., Sakamura, S., 1981. N-Carboxymethyl-l-serine a newacidi amino acid from Asparagus (Asparagus officinalis) shoots. Agricultural and Biological Chemistry 45, 1483-1485.

Kayaalp SO (1998). Medical pharmacology, in terms of rational treatment (Rasyonel tedavi yonunden tibbi farmakoloji), Ankara:

Hacettepe- Tas Ltd.Sti.

Khan N A and Rashid A. Z. M. M 2006. A Study on the Indigenous Medicinal Plants and Healing Practices in Chittagong Hill Tracts (Bangladesh). *Afr. J. Trad. CAM* (2006) 3 (3): 37 - 47

Khan N A and Rashid A. Z. M. M 2006. A Study on the Indigenous Medicinal Plants and Healing Practices in Chittagong Hill Tracts (Bangladesh). *Afr. J. Trad. CAM* (2006) 3 (3): 37 - 47

Kharkil M., Tiwari B., Bhadoni A., Bhatarai N. (2003). Creating Livelihoods Enhancing Medicinal and Aromatic Plants based Biodiversity-Rich Production Systems: Preliminary Lessons from South Asia. Oral paper presented at The 3rd World Congress on Medicinal and Aromatic Plants for Human Welfare (WOCMAP III). Chiang Mai, Thailand.

Kim DSHL, Park SY, Kim JY. Curcuminoids from Curcuma longa L. (Zingiberaceae) that protect PC12 rat pheochromocytoma and normal human umbilical vein endothelial cells from bA(1-42) insult. Neurosci Lett 2001;303(1):57- 61.

Kim SY, Moon TC, Chang HW, Son KH, Kang SS, Kim HP. Effects of tanshinone I isolated from Salvia miltiorrhiza Bunge on arachidonic acid metabolism and in vivo inflammatory responses. Phytother Res 2002;16(7):616- 20.

Kiritkar KR and Basu BD (1975) Indian Medicinal Plants, Indological and Oriental Publishers, Delhi, India.

Krishnaswamy M and Purushottamam KK (1980) Plumbagin, a study of its anticancer, antibacterial and antifungal properties, Ind. J. Exp. Biol. 18 : 876-877.

Kubmarawa, D., Ajoku, G.A., Enwerem, N.M. and Okorie, D.A. 2007. Preliminary phytochemical and antimicrobial screening of 50 medicinal plants from Nigeria, African J Biotechnology 6(14): 1690- 1696.

Kulkarni C, Pattanshetty JR, Amruthraj G. Effect of alcoholic extract of Clitoria ternatea Linn. on central nervous system in rodents. Indian J Exp Biol 1988;26:957-60.

Lake JT, Grossberg GT. Management of psychosis, agitation, and other behavioural problems in Alzheimer's disease. Psychiatr Ann 1987;17: 371- 4.

Lambert J., Srivastava, J., and Vietmeyer N. (1997). Medicinal plants : Rescuing a global heritage. World bank. Agriculture and Forestry Systems, Washington, D.C.

Lamia Sharmin (2004). Cultivation prospect of medicinal plants in Bangladesh: experiences from Natore (Unpublished Paper)

Lampronti I, Khan MTH, Borgatti M, Bianchi N and Gambari R. Inhibitory Effects of Bangladeshi Medicinal Plant Extracts on Interactions between Transcription Factors and Target DNA Sequences. *eCAM* 2008;5(3)303–312

Lau ST, Lin ZX, Liao Y, Zhao M, Cheng CHK, Leung PS: Brucein D induces apoptosis in pancreatic adenocarcinoma cell line PANC-1 through the activation of p38-mitogen activated protein kinase. Cancer Letters 2009, 281(1):42-52

Lebert F, Pasquier F, Petit H. Behavioural effects of trazodone in Alzheimer's disease. J Clin Psychiatry 1994;55(12):536- 8.

Lee MK, Kim SR, Sung SH, Lim DY, Kim H, Choi H, et al. Asiatic acid derivatives protect cultured neurons from glutamate-induced excitotoxicity. Res Commun Mol Pathol Pharmacol 2000;108(1–2):75- 86.

Lee S, Xiao C and Pei S 2008. Ethnobotanical survey of medicinal plants at periodic markets of Honghe Prefecture in Yunnan Province, SW China. *Journal of Ethnopharmacology* ; 117: 362-377.

Liao CH, Sang S, Liang YC, Ho CT, Lin JK. Suppression of inducible nitric oxide synthase and cyclooxygenase-2 in downregulating nuclear factor-kappaB pathway by Garcinol. Mol Carcinog 2004;41:140–9.

Liao CH, Sang S, Liang YC, Ho CT, Lin JK. Suppression of inducible nitric oxide synthase and cyclooxygenase-2 in downregulating nuclear factor-kappaB pathway by Garcinol. Mol. Carcinog 2004;41:140-9.

Lin, Y.S., Chiang, H.C., Kan, W.S., Hone, E., Shih, S.J., Won, M.H. 1992. The American Journal of Chinese Medicine 20, 233-243.

Loncar MB, Al-azzeh ED, Sommer PS, Marinovic M, Schmehl K, Kruschewski M, et al. Tumour necrosis factor alpha and nuclear factor kappaB inhibit transcription of human TFF3 encoding a gastrointestinal healing peptide. Gut 2003;52:1297-303.

Loncar MB, Al-azzeh ED, Sommer PS, Marinovic M, Schmehl K, Kruschewski M, et al. Tumour necrosis factor alpha and nuclear factor kappaB inhibit transcription of human TFF3 encoding a gastrointestinal healing peptide. Gut 2003;52:1297-303.

M, Rahmatullah 2009 An Ethnomedicinal Survey of Dhamrai Sub-district in Dhaka District, Bangladesh American-Eurasian Journal of Sustainable Agriculture, American-Eurasian Journal of Sustainable Agriculture, 4(2): 111-116, 2010

M, Rahmutullah, 2010 (a) A randomized survey of medicinal plants used by folk medicinal practitioners in six districts of Bangladesh to treat rheumatoid arthritis. *Advances in Natural and Applied Sciences > May 1, 2010*

M, Rahmutullah, 2010 (a) A randomized survey of medicinal plants used by folk medicinal practitioners in six districts of Bangladesh to treat rheumatoid arthritis. Advances in Natural and Applied Sciences. May 1, 2010

M. Parvez Rana, 2010 Ethno-medicinal plants use by the Manipuri tribal community in Bangladesh Journal of Forestry Research (2010) 21(1): 85-92.

M.R. Islam, 2009 In vitro Clonal Propagation of Vitex negundo L. - An Important Medicinal Plant Plant Tissue Cult. & Biotech. 19(1): 113-117, PTC&B

M.Zashim.Uddin ,et al,2010. Ethnobotanical Survey of Medicinal Plants in Phulbari Upazila of Dinajpur District , Bangladesh.

Bangladesh J. Plant Taxon. 13(1): 63-68, 2006 (June)

Mahesh B and Satish S. Antimicrobial Activity of Some Important Medicinal Plant Against Plant and Human Pathogens. *World Journal of Agricultural Sciences* 4 (S): 839-843, 2008

Manyam BV. Dementia in Ayurveda. J Altern Complement Med 1999; 5(1):81- 8.

Masuda A, Yoshikai Y, Kume H, Matsuguchi T. The interaction between GATA proteins and activator protein-1 promotes the transcription of IL-13 in mast cells. J Immunol 2004;173:5564 73.

Masuda A, Yoshikai Y, Kume H, Matsuguchi T. The interaction between GATA proteins and activator protein-1 promotes the transcription of IL-13 in mast cells. J Immunol 2004;173:5564 73.

Md. Rahman a Rahman, Use of Medicinal Plants by Folk Medicinal Practitioners among a Heterogeneous Population of Santals and Non-santals in Two Villages of Rangpur District, Bangladesh American-Eurasian Journal of Sustainable Agriculture, 4(2): 204-210, 2010 ISSN 1995-0748

Melanie-Jayne R. 2003, Plants used in Chinese and Indian traditional medicine for improvement of memory and cognitive function Pharmacology, Biochemistry and Behavior 75 (2003) 513–527.

Miquel J, Bernd A, Sempere JM, Diaz-Alperi J, Ramirez A. The curcuma antioxidants: pharmacological effects and prospects for future clinical use. A review. Arch Gerentol Geriatr 2002;34(1):37 –46.

Modi J (1961) Textbook of Medicinal Jurisprudence and toxicology, Pripati Pvt. Ltd.: Bombay, India.

Mohammed R, and M. A Haque Mollik, 2010, A Survey on the Use of Medicinal Plants by Folk Medicinal Practitioners in Five Villages of Boalia Sub-district, Rajshahi District, Bangladesh.

Advances in Natural and Applied Sciences, 4(1): 39-44,

Mohammed Rahmatullah, 2009 A survey of medicinal plants in two areas of Dinajpur district, Bangladesh including plants which can be used as functional foods American-Eurasian Journal of Sustainable Agriculture, 3(4): 862-876, 2009

Mollik MAH, Hossan MS, Paul A K, Rahman MT, Jahan R and Rahmatullah M 2010: A Comparative Analysis of Medicinal Plants Used by Folk Medicinal Healers in Three Districts of Bangladesh and Inquiry as to Mode of Selection of Medicinal Plants, Ethnobotany, *Research & Applications* 8:195-218.

Monteleone G, Mann J, Monteleone I, Vavassori P, Bremner R, Fantini M, et al. A failure of TGFbeta 1 negative regulation maintains sustained NF-KB activation in gut inflammation. J Biol Chem 2004;279:3925–32.

Monteleone G, Mann J, Monteleone I, Vavassori P, Bremner R, Fantini M, et al. A failure of TGFbeta 1 negative regulation maintains sustained NF-KB activation in gut inflammation. J Biol Chem 2004;279:3925–32.

Moon TC, Murakami M, Kudo I, Son KH, Kim HP, Kang SS, et al. A new class of COX-2 inhibitor, rutaecarpine from Evodia rutaecarpa. Inflamm Res 1999;48(12):621– 5.

Mukherjee PK, 2008 The Ayurvedic medicine Clitoria ternatea-- from traditional use to scientific assessment J Ethnopharmacol. 2008 Dec 8;120(3):291-301. Epub 2008 Sep 20.

Mukul SA, Uddin M B and Tito M R 2007: Medicinal Plant Diversity and Local Healthcare Among The People Living in and Around a Conservation area of Northern Bangladesh, *Int. J. For. Usuf. Mngt.* 8 (2) : 50-63.

Naik GH, Priyadarsini KI, Naik DB, Satav JG, Mohan H. Antioxidant activity and phytochemical analysis of the aqueous extract of Terminalia chebula. Free Radic Biol Med 2002;33(Suppl 1):547.

Nakajima T, Aratani S, Nakazawa M, Hirose T, Fujita H, Nishioka K. Implications of transcriptional coactivator CREB binding protein complexes in rheumatoid arthritis. Mod Rheumatol 2004;14:6-11.

Nakajima T, Aratani S, Nakazawa M, Hirose T, Fujita H, Nishioka K. Implications of transcriptional coactivator CREB binding protein complexes in rheumatoid arthritis. Mod Rheumatol 2004;14:6-11.

Nalini K, Aroor AR, Karanth KS, Rao A. Effect of Centella asiatica fresh leaf aqueous extract on learning and memory and biogenic amine turnover in albino rats. Fitoterapia 1992;63(3):232- 7.

Nozaki K, Hikiami H, Goto H, Nakagawa T, Shibahara N, Shimada Y. Keishibukuryogan (Gui-Zhi-Fu-Ling-Wan), a Kampo Formula, Decreases Disease Activity and Soluble Vascular Adhesion Molecule-1 in Patients with Rheumatoid Arthritis. Evid Based Complement Alternat Med 2006;3:359-64.

Nozaki K, Hikiami H, Goto H, Nakagawa T, Shibahara N, Shimada Y. Keishibukuryogan (gui-zhi-fu-ling-wan), a kampo formula, decreases disease activity and soluble vascular adhesion molecule-1 in patients with rheumatoid arthritis. Evid Based Complement Alternat Med 2006;3:359-64.

Nozaki K, Hikiami H, Goto H, Nakagawa T, Shibahara N, Shimada Y. Keishibukuryogan (Gui-Zhi-Fu-Ling-Wan), a Kampo Formula, Decreases Disease Activity and Soluble Vascular Adhesion Molecule-1 in Patients with Rheumatoid Arthritis. Evid Based Complement Alternat Med 2006;3:359-64.

Nozaki K, Hikiami H, Goto H, Nakagawa T, Shibahara N, Shimada Y. Keishibukuryogan (gui-zhi-fu-ling-wan), a kampo formula, decreases disease activity and soluble vascular adhesion molecule-1 in patients with rheumatoid arthritis. Evid Based Complement Alternat Med 2006;3:359-64.

Oliver, B. (1960). Medicinal plants in Nigeria, Nigerian College of Arts, Science and Technology, Lagos.

Panda BB, Gaur K, Nema RK, Sharma CS, Jain AK, Jain CP (2009). Hepatoprotective activity of *Jatropha gossypifolia* against carbon tetrachloride- induced hepatic injury in rats. Asian Gossypiline, a new lignan from *Jatropha*. J. Pharm. Clin. Res. 2: 50-54.

Pandey K., Sinha P.K., Das V. R., Bimal S., Singh S. K. & Das P. (2009) Pharmacotherapeutic options for Visceral leishmaniasis – Current scenario Clinical Medicine: Pathology Clinical Medicine: Pathology 2 :1-4.

Park CH, Choi SH, Koo JW, Seo JH, Kim HS, Jeong SJ, et al. Novel cognitive improving and neuroprotective activities of Polygala tenuifolia Willdenow extract, BT-11. J Neurosci Res 2002;70(3):484-92.

Parrotta, J.A., 2001. The Healing Plants of Peninsular India. MRM Graphics Ltd., Winslow, Bucks.

Patil HM and Bhaskar VV 2006: Medicinal Uses of Plants by Tribal Medicine Men of Nandurbar district in Maharashtra, *Natural Product Radiance* 5(2): 125- 130.

Penso G.1980. WHO inventory of medicinal plants used in different countries. - Geneva, Switzerland;

Perry, L.M., 1980.Medicinal Plants of East and South-East Asia— Attributed Properties and Uses. MIT Press, Cambridge (United States) and London.

Pezzuto JM: Plant-derived anticancer agents. Biochemical pharmacology 1997, 53(2):121-133.

Pezzuto JM: Plant-derived anticancer agents. Biochemical pharmacology 1997, 53(2):121-133.

Pillai NGK, Menon TV, Pillai GB, Rajasekharan S and Nair CRR (1981) Effect of plumbagin in Charmakeela (common warts) a case report. J. Res. Ayur. Sidha 2 : 12-126.

Premakumari P, Rathinam K and Santhakumari G (1977) Antifertility activity of plumbagin. Ind. J. Med. Res. 65 : 829-838.

Priyadarsini Kl. Free radical reactions of curcumin in membrane models. Free Radic Biol Med 1997;23(6):838- 43.

Rahman AHMM, Anisuzzaman M, Haider SA, Ahmed F, Islam AKMR and Naderuzzaman ATM 2008: Study of Medicinal Plants in the Graveyards of Rajshahi City, *Research Journal of Agriculture and Biological Sciences* 4(1): 70-74.

Rahman AHMM, Anisuzzaman M, Haider SA, Ahmed F, Islam AKMR and Naderuzzaman ATM 2008: Study of Medicinal Plants in the Graveyards of Rajshahi City, *Research Journal of Agriculture and Biological Sciences* 4(1): 70-74.

Rahman AK, Chowdhury AK, Ali HA, Raihan SZ, Ali MS, Nahar I, et al. (2009) Antibacterial activity of two limonoids from Swietenia mahagoni against multiple-drug-resistant (MDR) bacterial strains. Nat Med (Tokyo); 63:41-5.

Rahman M A, Islam S, Naim N, M H. Chowdhury M H, Jahan R, Rahmatullah M 2010: Use of Medicinal Plants by Folk Medicinal Practitioners among a Heterogeneous Population of Santals and Non-santals in Two Villages of Rangpur District, *Bangladesh American-Eurasian Journal of Sustainable Agriculture*, 4(2): 204-210.

Rahman M S, Rahman M Z, Wahab M A , Chowdhury R and Rashid M A. Antimicrobial Activity of Some Indigenous Plants of Bangladesh. *Dhaka Univ. J. Pharm. Sci.* 7(1): 23-26, 2008 (June)

Rahman M.S., Sadhu S.K. and Hasan C.M. (2007) Preliminary antinociceptive, antioxidant and cytotoxic activities of Leucas aspera root, Fitoterapia 78: 552–555.

Rahmatullah M, Azam MN K, Mollik MAH, Hasan MM , Hassan AI, Jahan R , Jamal F, Nasrin D, Ahmed R, Rahman MM, Khatun MA 2010b: Medicinal plants used by the Kavirajes of Daulatdia Ghat, Kushtia district, Bangladesh . *American-Eurasian Journal of Sustainable Agriculture*, 4(2): 219-229.

Rahmatullah M, Mollik M AH, Khatun MA ,R Jahan R, Chowdhury A R, Seraj S, Hossain M S, Nasrin D, Khatun Z 2010e. A Survey on the Use of Medicinal Plants by Folk Medicinal Practitioners in Five Villages of Boalia Sub-district, Rajshahi District, Bangladesh. *Advances in Natural and Applied Sciences*, 4(1): 39-44,

Rahmatullah M, Nuruzzaman, Hossan S , Khatun A , Rahman M, Jamal F 2010c: An ethnomedicinal survey of folk medicinal practitioners of Shitol Para village, Jhalokati district, Bangladesh. *Advances in Natural and Applied Sciences,* 4(1): 85-92, 201.

Rahmatullah M., 2010 (d) A Survey of Medicinal Plants used by Folk Medicinal Practitioners in Balidha village of Jessore District, Bangladesh American-Eurasian Journal of Sustainable Agriculture, 4(2): 111-116, 2010 ISSN 1995-0748

Rai KS, Murthy KD, Karanth KS, Nalini K, Rao MS, Srinivasan KK. Clitoria ternatea root extract enhances acetylcholine content in rat hippocampus. Fitoterapia 2002;73(7 – 8):685–9.

Rajakrishnan V, Viswanathan P, Rajasekharan KN, Menon VP. Neuroprotective role of curcumin from Curcuma longa on ethanol-induced brain damage. Phytother Res 1999;13(7):571–4.

Ramsewak RS, DeWitt DL, Nair MG. Cytotoxicity, antioxidant and antiinflammatory activities of curcumins I –III from Curcuma longa. Phytomedicine 2000;7(4):303 – 8.

Rana M P, Sohel M S I, Akhter S, Islam MJ 2010. Ethno-medicinal plants use by the Manipuri tribal community in Bangladesh. *Journal of Forestry Research* 21(1):85-92.
Rao VR and Arora RK. Rationale for conservation of medicinal plants. In: Batugal PA et al. (eds), Medicinal Plants Research in Asia. Vol.1:the Framework and Project Work Plans. International Plant Genetic Resources Institute (IPGRI), KualaLumpur, Malaysia; 2004:7–22.

Reinikainen KJ, Paljarvi L, Halonen T, Malminen O, Kosma V-M, Laakso M, et al. Dopaminergic system and monoamine oxidase-B activity in Alzheimer's disease. Neurobiol Aging 1988;9:245 –

52.

Reisberg B, Ferris S, Mobius HJ, Schmitt F, Doody R. Long-term treatment with the NMDA antagonist memantine: results of a 24-week, open-label extension study in moderately severe-to-severe Alzheimer's disease. Neurobiol Aging 2002;23(Suppl 1):2039.

Ribeiro A, Romeiras MM, Tavares J, Faria MT. Ethnobotanical survey in Canhane village, district of Massingir, Mozambique: medicinal plants and traditional knowledge. J Ethnobiol Ethnomed. 2010 Dec 3;6:33. doi: 10.1186/1746-4269-6-33.
Saeed S, Tariq P 2006: Pharmacological activities of ginger (*Zingiber officinale*). *International Journal of Biology and Biotechnology* 3 519-526.

Saha M. R., Jahangir R., Vhuiyan M.. M. I. and Biva I. J. (2008). In Vitro Nitric Oxide Scavenging Activity of Ethanol Leaf Extracts of Four Bangladeshi Medicinal Plants, *S. J. Pharm. Sci.* 1(1&2): 57-62.

Sakarkar DM, Gonsalvis L, Fatima F, Khandelwal LK, Jaiswal B, Pardeshi MD 2006: Turmeric: an excellent traditional herb. *Plant Archives* 6 451-458.

Salem ML. Immunomodulatory and therapeutic properties of the Nigella sativa L. seed. Int Immunopharmacol 2005;5:1749-70.

Salem ML. Immunomodulatory and therapeutic properties of the Nigella sativa L. seed. Int Immunopharmacol 2005;5:1749-70.

Sato K, Takayanagi H. Osteoclasts, rheumatoid arthritis, and osteoimmunology. Curr Opin Rheumatol 2006;18:419-26.

Sato K, Takayanagi H. Osteoclasts, rheumatoid arthritis, and osteoimmunology. Curr Opin Rheumatol 2006;18:419-26.

Scartezzini P, Speroni E. Review on some plants of Indian traditional medicine with antioxidant activity. J Ethnopharmacol 2000;71(1-2): 23- 43.

Schneider, W., 1985. Wörterbuch der Pharmazie, 4 Geschichte

der Pharmazie. Wissenschaftliche Verlagsgesellschaft mbH Stuttgart, p. 127.

Schwarz RE, Donohue CA, Sadava D, Kane SE: Pancreatic cancer in vitro toxicity mediated by Chinese herbs SPES and PC-SPES: implications for monotherapy and combination treatment. Cancer Letters 2003, 189(1):59-68.

Sebban H, Courtois G. NF-kappaB and inflammation in genetic disease. Biochem Pharmacol 2006;72:1153–60.

Sebban H, Courtois G. NF-kappaB and inflammation in genetic disease. Biochem Pharmacol 2006;72:1153–60.

Seidl R, Cairns N, Singewald N, Kaehler ST, Lubec G. Differences between GABA levels in Alzheimer's disease and down syndrome with Alzheimer- like neuropathology. Naunyn Schmiedebergs Arch Pharmacol 2001;363(2):139- 45.

Sen A. , Sharma M.M.,. Grover D and. Batra A .In Vitro Regeneration of Phyllanthus amarus Schum. And Thonn.: An Important Medicinal Plant .Our Nature (2009) 7:110-115

Shahwar D., Rehman S.U., Ahmad N., Ullah S. and Raza M. A. (2010) Antioxidant activities of the selected plants from the family Euphorbiaceae, Lauraceae, Malvaceae and Balsaminaceae. African Journal of Biotechnology 9(7): 1086-1096.

Shanley P and Luz L 2003. The impacts of forest degradation on medicinal plant use and implications for health care in eastern Amazonia. *BioScience* ; 53 (6): 573-584.

Shao,Y., Poobrasert, O.,Kennelly, E.J., Chin, C.K., Ho, C.T., 1997. Steroidal saponins from Asparagus officinalis and their xytotoxic activity. Planta Medica 63, 258–262.

Shukla PK, Khanna VK, Ali MM, Maurya RR, Handa SS, Srimal RC. Protective effect of Acorus calamus against acrylamide induced neurotoxicity. Phytother Res 2002;16(3):256 –60.

Simpson PM, Foster D. Improvement in organically disturbed

behavior with trazodone treatment. J Clin Psychiatry 1986;47:191-3.

SK Kulkarni - 2008 - Withania somnifera: An Indian ginseng Progress in Neuro-Psychopharmacology and Biological Psychiatry • Volume 32, Issue

Skrzypczak-Jankun E, McCabe NP, Selman SH, Jankun J. Curcumin inhibits lipoxygenase by binding to its central cavity: theoretical and Xray evidence. Int J Mol Med 2000;6(5):521 -6.

Soares, M.B.P., Bellintani, M.C., Ribeiro, I.M., Tomassini, T.C.B., Santos, R.R., 2003. Inhibition of macrophage activation and lipopolysaccharide-induced death by seco-steroids purified from Physalis angulata L. European Journal of Pharmacology 459, 107–112.

Sofowora, A., 1982. Medicinal Plants and Traditional Medicinal in Africa. John Wiley and Sons, New York, pp: 256.

Sosa S, Balick MJ, Arvigo R, Esposito RG, Pizza C, Altinier G (2002). A Screening of the American plants. J. Ethanopharmacol. 8: 211-215.

Srivastava KC, Bordia A, Verma SK. Curcumin, a major component of food spice tumeric (Curcuma longa) inhibits aggregation and alters eicosanoid metabolism in human blood platelets. Prostaglandins Leukot Essent Fatty Acids 1995;52(4):223 -7.

Srivastava, J., J. Lambert and N. Vietmeyer, 1996. Medicinal plants: An expanding role in development. J.World Bank Technical Paper. No. 320.

Srivastava, J., Lambert J. and Vietmeyer N. (1996). Medicinal plants : An expanding role in development. World bank. Agriculture and Forestry Systems, Washington, D.C.

Stone, R., Xin, H., 2006. Novartis invests $100 million in Shanghai. Science 314, 1064–1065.

Storga D, Vrecko K, Birkmayer JG, Reibnegger G.

Monoaminergic neurotransmitters, their precursors and metabolites in brains of Alzheimer patients. Neurosci Lett 1996;203(1):29- 32.

Stucchi A, Reed K, O'Brien M, Cerda S, Andrews C, Gower A, et al. A new transcription factor that regulates TNF-alpha gene expression, LITAF, is increased in intestinal tissues from patients with CD and UC. Inflamm Bowel Dis 2006;12:581-7.

Stucchi A, Reed K, O'Brien M, Cerda S, Andrews C, Gower A, et al. A new transcription factor that regulates TNF-alpha gene expression, LITAF, is increased in intestinal tissues from patients with CD and UC. Inflamm Bowel Dis 2006;12:581-7.

SyedWaseem Bihaqia,2009 Neuroprotective role of Convolvulus pluricaulis on aluminium induced neurotoxicity in rat brain Journal of Ethnopharmacology 124 (2009) 409-415

Taranalli AD, Cheeramkuzhy TC. Influence of Clitoria ternatea extracts on memory and central cholinergic activity in rats. Pharm Biol 2000;38(1): 51-6.

Tohda C, Nakayama N, Hatanaka F, Komatsu K. Comparison of Anti-inflammatory Activities of Six Curcuma Rhizomes: a Possible Curcuminoid-independent Pathway Mediated by Curcuma phaeocaulis Extract. Evid Based Complement Alternat Med 2006;3:255-60.

Tohda C, Nakayama N, Hatanaka F, Komatsu K. Comparison of Anti-inflammatory Activities of Six Curcuma Rhizomes: a Possible Curcuminoid-independent Pathway Mediated by Curcuma phaeocaulis Extract. Evid Based Complement Alternat Med 2006;3:255-60.

Uchihara T, Kondo H, Kosaka K, Tsukagoshi H. Selective loss of nigral neurons in Alzheimer's disease: a morphometric study. Acta Neuropathol 1992;83:271- 6.

Uddin MZ, Hassan MA and Sultana M 2006: Ethhnobotanical Survey of Medicinal Plants in Phulbari Upazilla of Dinajpur District, Bangladesh, Bangladesh J. Plant Taxon 13(1): 63-68.

Uddin MZ, Hassan MA and Sultana M 2006: Ethhnobotanical Survey of Medicinal Plants in Phulbari Upazilla of Dinajpur District, Bangladesh, *Bangladesh J. Plant Taxon* 13(1): 63-68.
Uddin S J, Grice ID and Tiralongo E, 2009. Cytotoxic Effects of Bangladeshi Medicinal Plant Extracts. *eCAM 2009;*Page 1 of 6

Ukiya, M., Akihisa, T., Yasukawa, K., Toduda, H., Toriumi, M., Koike K., et al., 2002. Anti-inflammatory and antitumor-promoting effects of cucurbitane glycosides from the roots of Bryonia dioica. Journal of Natural Products 65, 179–183.

Uniyal, S.K., K.N. Singh, P. Jamwal and B. Lal, 2006. Traditional use of medicinal plants among the tribal.communities of Chhota Bhangal, Western Himalayan. J. Ethnobiol. Ethnomed., 2: 1-14.

Vankar P.S., Tiwari1 V., Singh L.W. and Swapana N. (2006). Antioxidant Properties of some exclusive species of Zingiberaceae family of Manipur. *EJEAFChe*, 5 (2):1318-1322

Vasishta, P.C. 1974. Taxonomy of Angiosperms 2nd ed., R. Chand & Co., New Delhi.

Vasudevan M, Parle M 2007: Memory enhancing activity of Anwala chuma (*Emblica officinalis*): an Ayurvedic preparation. *Physiology and Behavior* 91 46-54.

Ven Murthy MR , 2010.Scientific basis for the use of Indian ayurvedic medicinal plants in the treatment of neurodegenerative disorders: ashwagandha. Cent Nerv Syst Agents Med Chem. 2010 Sep 1;10(3):238-46

Verpoorte R. 2007. Bioprospectin: exploring our only renewable natural resorce. XXVII Annual Meeting on Micromolecular Evolution, Systematics and Ecology Reflections on the Current Status of Chemosystematics. BPL-3. São Paulo, Brasil.

Vikas Kumar 2006, Potential Medicinal Plants for CNS Disorders: an Overview Phytother. Res. 20, 1023–1035

Vlachojannis, J., B.D. Roufogalis and S. Chrubasik, 2008. Systematic review on the safety of Harpagophytum preparations for osteoarthritic and low back pain. Phytotherapy

Research, 22:149-152.

Vohora SB, Shah SA, Dandiya PC. Central nervous system studies on an ethanol extract of Acorus calamus rhizomes. J Ethnopharmacol 1990; 28:53- 62.

Vohra, B.P.S., Gupta, S.K., 2005. Indian ayurvedic medicine in aging prevention and treatment. Aging Interventions and Therapies, 303-327.

Wagner EF, Eferl R. Fos/AP-1 proteins in bone and the immune system. Immunol Rev 2005;208:126-40.

Wagner EF, Eferl R. Fos/AP-1 proteins in bone and the immune system. Immunol Rev 2005;208:126-40.

Walter S. Non-Wood Forest Products in Africa. A Regionaland National Overview. FAO, Rome, Italy; 2001.

Warnock, M., D. McBean, A. Suter, J. Tan and P. Whittaker, 2007. Effectiveness and safety of Devil's Claw tablets in patients with general rheumatic disorders. Phytotherapy Research, 21: 1228-1233.

Warrier PK, Nambiar VPK, Ramankutty C. Indian medicinal plants, vol. 2. India: Orient Longman; 1995.

Wasfi, I.A., Bashir, A.K., Abdalla, A.A., Banna, N.R., Tanira, M.O.M., 1995. Antiinflammatory activity of some medicinal plants of the United Arab Emirates. International Journal of Pharmacology 33, 124-128.

Wenzig, E., Kunert, O., Ferreira, D., Schmid, M., Schühly, W., Baurer, R., Hiermann, A., 2005. Flavanolignans from Avena sativa. Journal of Natural Products 68, 289-292.

White, N.J., 2008. Qinghaosu (Artemisinin): the price of success. Science 320,330-334.

Williamson, M.E., 2002. Major Herbs of Ayurveda. Churchill Livingston, London, UK.

Winblad B, Poritis N. Memantine in severe dementia: results of the 9MBEST study (benefit and efficacy in severely demented patients during treatment with mementine). Int J Geriatr Psychiatry 1999;14:135- 46

Wolters, B., 1994. Drogen, Pfeilgift und Indianermedizin: Arzneipflanzen aus S¨udamerika. Urs Freud Verlag GmbH, Greifenberg, Germany.

Wu MJ, Wang L, Ding HY, Weng CY, Yen JH. Glossogyne tenuifolia acts to inhibit inflammatory mediator production in a macrophage cell line by downregulating LPS-induced NF-kappaB. J Biomed Sci 2004;11:186–99.

Wu MJ, Wang L, Ding HY, Weng CY, Yen JH. Glossogyne tenuifolia acts to inhibit inflammatory mediator production in a macrophage cell line by downregulating LPS-induced NF-kappaB. J Biomed Sci 2004;11:186–99.

Wu TL and Larsen K (2000) Zingiberaceae. Flora of China 24:322–377

Wu, S.J., Tsai, J.Y., Chang, S.P., Lin, D.L., Wang, S.S., Huang, S.N., Ng, L.T., 2006. Supercritical carbon dioxide extract exhibits enhanced antioxidant and antiinflammatory activities of Physalis peruviana. Journal of Ethnopharmacology 108, 407–413.

www.herbpalace.com

www.naturalhealthschool.com

Yasmin H, Kaisar MA, Sarker MMR, Rahman MS and.Rashid M A Preliminary Anti-bacterial Activity of Some Indigenous Plants of Bangladesh. *Dhaka Univ. J. Pharm. Sci.* 8(1): 61-65, 2009 (June)

Yoshimura A, Mori H, Ohishi M, Aki D, Hanada T. Negative regulation of cytokine signaling influences inflammation. Curr Opin Immunol 2003;15:704-8.

Yoshimura A, Mori H, Ohishi M, Aki D, Hanada T. Negative regulation of cytokine signaling influences inflammation. Curr

Opin Immunol 2003;15:704–8.

Yu ZF, Kong LD, Chen Y. Antidepressant activity of aqueous extracts of Curcuma longa in mice. J Ethnopharmacol 2002;83(1- 2):161–5.

Zanoli P, Avallone R, Baraldi M. Sedative and hypothermic effects induced by b-asarone, a main component of Acorus calamus. Phytother Res 1998;12(Suppl 1):S114–6.

ABOUT THE AUTHOR

Mst Monira Khaton studied Biotechnology and Genetic Engineering in Islamic University, Kushtia, Bangladesh. She also worked in crop Science in University of Padova. She is working as Research Asst. in Dept. of Biotechnology, Bangladesh Agricultural University. Mst. Khaton also published couples of Research articles in Scientific Journal. She is interested on the Medicinal Plants in Bangladesh and traditional ways of diseases treatment.

Dr. Md Munan Shaik, is a Assistant Professor at Dept. of Biotechnology and Genetic Engineering,Islamic University, Bangladesh. He has received PhD from University of Padova, Italy and had Postdoctoral Training from IBS, CNRS, France.